Waste Management for a Sustainable Future - Technologies, Strategies and Global Perspectives

Edited by Hosam M. Saleh and Amal I. Hassan

Published in London, United Kingdom

Waste Management for a Sustainable Future - Technologies, Strategies and Global Perspectives
http://dx.doi.org/10.5772/intechopen.1004499
Edited by Hosam M. Saleh and Amal I. Hassan

Contributors
Andrea Colantoni, Anusha Kolanji, Awadhesh Kumar, Azizan Marzuki, Bárbara Abreu, Davor Dolar, Dharunya Rajagopal, Dong-Cheon Seo, Dong-Hoon Lee, Enrico Maria Mosconi, G.M. Golam Mehebub Pantho, Henrique Costa, Hubert Boeckmann, Ivan Brandić, Iva Ćurić, Kalaivani Ramanathan, Kokila Kannusamy, Kyeong-Sik Kim, Luka Brezinščak, Marcelo Oliveira, Mariarita Tarantino, Mohammad Borhan Uddin, Mohana Priya Rajendren, Muhammad Saleh, Murugesan Manikkampatti Palanisamy, Muthuminal Ramu, Narendra Kumar Rana, Oyebola Odunayo Olabinjo, Padmapriya Veerappan, Ratin Al Samir, Sankaranarayanan Devikarani, Saranya Indrabal, Soonkeum Han, Yong-Woo Jeon, Zezhong Wang

First published in London, United Kingdom, 2025 by IntechOpen
IntechOpen is the global imprint of INTECHOPEN LIMITED, registered in England and Wales, registration number: 11086078, 167-169 Great Portland Street, London, W1W 5PF, United Kingdom

For EU product safety concerns: IN TECH d.o.o., Prolaz Marije Krucifikse Kozulić 3, 51000 Rijeka, Croatia, info@intechopen.com or visit our website at intechopen.com.

British Library Cataloguing-in-Publication Data
A catalogue record for this book is available from the British Library

Waste Management for a Sustainable Future - Technologies, Strategies and Global Perspectives
Edited by Hosam M. Saleh and Amal I. Hassan
p. cm.
Print ISBN 978-0-85014-778-0
Online ISBN 978-0-85014-777-3
eBook (PDF) ISBN 978-0-85014-779-7

Meet the editors

Hosam Saleh is a Professor of Radioactive Waste Management in the Radioisotope Department at the Atomic Energy Authority, Egypt. He was awarded an MSc and Ph.D. in Physical Chemistry from Cairo University. Saleh has more than 27 years of experience in hazardous waste management, emphasizing treatment and developing new matrixes for immobilizing these wastes. He is also interested in studying innovative economic and environment-friendly techniques for managing hazardous and radioactive wastes. He authored many peer-reviewed scientific papers and chapters and served as an editor of several books. He has been selected among the top 2% of scientists in the world according to the Stanford University reports published by Elsevier for 2019, 2020, 2021, 2022 and 2023.

Amal I. Hassan is a Professor of Animal Physiology in the Department of Radioisotopes, Nuclear Research Center at the Atomic Energy Authority in Egypt. She has authored many peer-reviewed papers on chronic disease problems. She is a referee, reviewer, and editorial board member for several international scientific journals. She received a Certificate of Excellence in international scientific research arbitration from Publons. She was also selected for inclusion in Who's Who in 2014, 2015, and 2016.

Contents

Preface

Today, the world faces a big problem with too much solid waste, which affects cities, rural areas, and industries everywhere. The fast growth of cities, increasing population, new technologies, and changing consumption habits are causing massive waste increases. As we keep developing, the need for better, efficient, and sustainable waste management systems is more important now than ever. This book looks at these urgent issues, providing an overview of modern waste management methods, new trends, and examples of success from around the world. Waste management is no longer simply about collection and disposal. It has evolved into a multi-disciplinary field integrating environmental science, engineering, public policy, technology, and social behavior. The chapters in this book explore the latest innovations in waste-to-energy technologies, recycling systems, and waste recovery practices. From cutting-edge techniques such as bioenergy with carbon capture and storage (BECCS) to intelligent non-destructive testing for industrial waste, this work offers insights into the technologies shaping the future of waste management.

The book is split into different themes, each showing an important part of waste management. The first theme looks at technology, especially waste-to-energy and bio-waste recovery. The second theme discusses industrial waste management, focusing on the environmental effects of harmful materials like chromium waste and lithium-ion batteries. Much of this book also looks at issues like microplastics, e-waste, and pharmaceutical waste, stressing the urgent need for better disposal and recycling methods.

At the core of waste management practices is municipal solid waste (MSW), and the book devotes several chapters to discussing best practices, policies, and successful case studies of effective MSW management in cities worldwide. Waste management strategies are also given special attention, including source segregation, community-based solutions, and public awareness. The book also discusses reusing waste in agriculture and other sectors, offering a vision of a circular economy that recycles waste instead of treating it as waste.

This book is based on a singular element of realizing waste management is also social, political and economic and highlights the recognition of networked approaches across sectors and scales.

The authors, who are skilled in their areas, give valuable insights, studies, and real-world solutions that may fit in different settings—from tall city buildings to countryside locations. Their work highlights the need to connect waste management with sustainable development aims, environmental protection, and public health.

This book is aimed primarily at students, researchers, policymakers, environmental professionals, and industry practitioners. Whether you are searching for cutting-edge technological solutions, policy frameworks, or real-world case studies, there is something here for everyone to help inform the future of waste management. As we head towards a more sustainable age, the management of solid waste will become imperative for the health of our planet. Because of innovative solutions, informed practices, and collaborative effort, waste can become a resource instead of a burden, and the foundation of a more sustainable and resilient future can be set. We hope this book can assist with this common struggle and ignite new ideas, initiatives, and actions to manage solid waste in ways that benefit not only people but also the planet.

Hosam M. Saleh and Amal I. Hassan
Egyptian Atomic Energy Authority,
Cairo, Egypt

Section 1

Waste-to-Energy, Innovations, and Technological Solutions

Chapter 1

A Quantum Leap for Waste and BECCS with B-VM Power Plant

Hubert Boeckmann

Abstract

This chapter details the development of a so-called B-VM power plant from a patented bioenergy process B-V. The objective is the valorization of municipal waste and the implementation of an upstream waste system for the B-V to process the delivered waste. Intermediate products of the B-VM process are storable pyrolysis coke, pyrolysis oil and synthesis gas which generate electrical and thermal energy. A portion of this energy is used to extract carbon dioxide from the process emissions for realizing BECCS (Bioenergy with Carbon Capture and Storage). Compared to the combustion of waste or biomass, the following ecological and economic benefits and innovations are achieved: Optimal valorization of residual waste and biomass; decoupling of thermal and electrical energy production; doubling of the net electric efficiency of the power plant; potential to achieve negative emissions through BECCS; capable for base and peak load and also very strong grid balancing.

Keywords: BECCS, waste utilization, waste pyrolysis, pyrolysis oil, 2-stroke ship diesel engine, energy decoupling, electrical efficiency, negative emissions, grid balancing

1. Introduction

1.1 The global problems of carbon dioxide (= CO_2) and waste

In addition to the widely recognized worldwide problem of CO_2 and its associated impact on the climate, waste pollution is also a significant global challenge. The issue of CO_2 is increasingly highlighted by the media and politicians as a global problem and is becoming more apparent due to the resulting restrictions, while waste primarily affects the local environment. However, the volume of waste continues to rise and appears increasingly threatening, as the data in the World Bank's "What a Waste" study [1] shows.

There are no quick solutions for reducing CO_2 emissions and significant investments and restrictions are required. In contrast, waste issues might be promisingly addressed through stringent management and the combustion or similar treatment of waste. The volume of global waste is expected to increase from 2016 by 1.7 times to approximately 3.4 billion tons annually in 2050. This escalation adversely affects populations and environments and leads to a widespread scarcity of land. An alternative and far superior method of waste disposal is undoubtedly thermal waste treatment. However, the waste incineration technologies available in the market are profoundly imperfect. These technologies are ineffective, inefficient, pollute the environment and almost invariably provoke societal controversies.

To combat the worsening waste problems and reduce the concentration of CO_2 in the atmosphere, a new process has been developed for power plants based on waste. Through this, the high proportion of biomass in the waste can also be used via BECCS. This innovative power plant concept has significantly better ecological and economic parameters than any waste incineration. The innovative approach and main difference to conventional waste incineration plants is that the underlying process uses pyrolysis instead of incineration, which determines the entire subsequent process, including a very special but world-renowned diesel engine.

As a result, improvements and new applications for waste management and BECCS can be realized, which may certainly be considered a quantum leap. It is noteworthy that almost all essential facts and circumstances have been long known and utilized, yet they had not been rethought or linked to a new solution until now.

1.2 Given project circumstances

The newly developed power plant for biomass and waste utilizes the so-called KSW Process® for biomass [2–4], which was developed and patented in Germany approximately 10 years ago (hereafter referred to as the B-V bioenergy process). Power plants employing this process harness the pyrolysis of the input for thermal valorization and produce storable energy sources in the form of pyrolysis coke and pyrolysis oil as intermediate products. These intermediates are utilized as synthesis gas and pyrolysis oil in slow-running, two-stroke, dual-fuel large diesel engines,

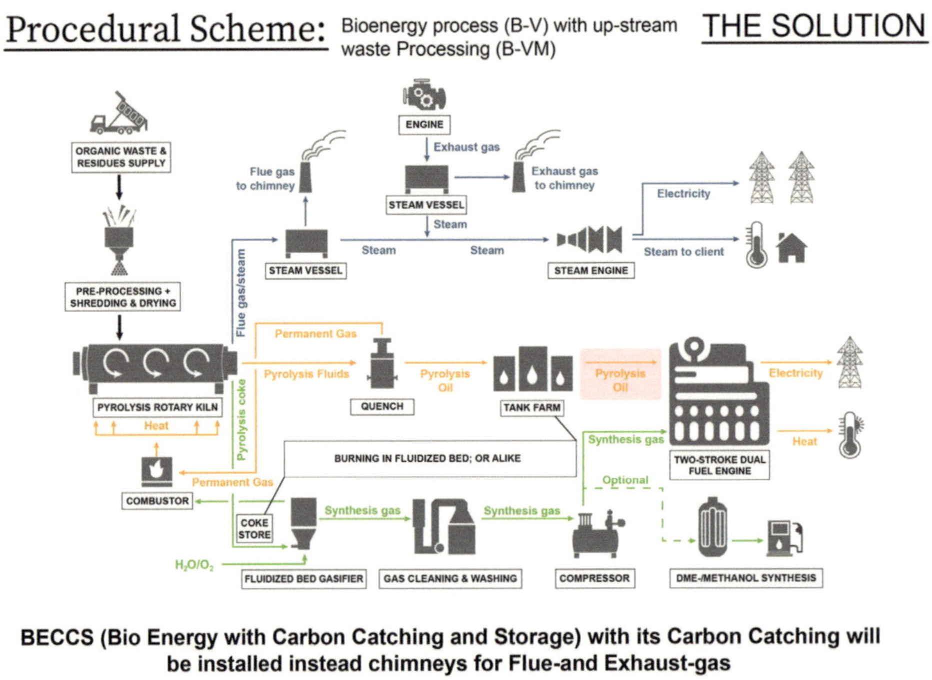

Figure 1.
New developed power plant for biomass and residual waste. From author's BMC Sp.z o.o. working group B-VM.

with generator capacities ranging from 3 to 80 MW. This concept leads to optimal ecological and economic performance metrics and offers numerous advantages. On this basis, a power plant with a 14 MW engine was designed, utilizing biomass and meeting the requirements of Germany's Renewable Energy Sources Act (EEG). The author of this chapter was involved in the economic assessment, the business plan and several technical variants.

These circumstances led to the idea of using municipal solid waste (in the broadest sense and almost always with very high biomass content) instead of pure biomass and implementing this as a separate project through a consortium in Poland. The primary decision was to employ the pyrolysis process to decompose plastics in the waste along with organic biomass. This includes plastics that would otherwise not be recycled and whose complete separation from waste and its fractions is either not feasible or only possible with great effort. This means, among other things, that the extraction of plastic fractions from the delivered waste can be realized at any time, provided there is an economic rationale or legal mandate.

Municipal solid waste requires appropriate processing as an optimal basis for the B-V pyrolysis, along with various adaptations within the existing B-V process. **Figure 1** illustrates the simplified process of a B-V with an upstream waste treatment, here referred to as B-VM.

Based on this, the five innovations mentioned in the abstract are presented in the following Sections 2–6 and summarized in a brief conclusion in Section 7.

2. Optimal residual waste utilization and pyrolysis

The high flexibility of pyrolysis in rotary kilns concerning operational parameters and tolerance to varying properties of feedstock is proving to be essential. This adaptability facilitates the processing of nearly all thermally degradable wastes or residues with minimal issues. A high carbon content of the input has a positive effect on the process with regard to profitability. Moreover, it is fundamentally important to understand the interactions between the input and the output of the pyrolysis to direct the output into desired fractions through targeted modifications of the input, thus optimizing the effectiveness and purposefulness of the process. The pyrolysis input must be prepared according to the required, optimal process conditions, ensuring that:

- the particle size is $\leq$20 mm,
- the residual moisture content is $\leq$10% and
- (nearly) all inert and valuable substances are extracted.

The use of B-V with more or less pure biomass leads to drying and crushing and possibly a system against metals and stones or similar. The entire processing is straightforward and therefore fully integrated into B-V.

However, when using municipal solid waste, waste processing becomes a distinct area before B-V and with the great variety of globally produced municipal waste, the systems required for the waste treatment are determined. All necessary technologies are well known and are being developed continuously. B-VM has several own patents in preparation. Sorted materials can be recycled, further processed, or secured. This enables optimal and sustainable implementation of locally and legally prescribed waste processing worldwide and, if necessary, later adaptation.

Typically, this range results in an output where, excluding the ≤10% residual moisture and ≥10% ash (also containing remaining inert substances), about 70–80% biomass (always +/– contaminated in waste) plus 30–20% non-recyclable plastics are fed into the process. This is based on dry material and the additional use of some regional biomass increases the corresponding share of biomass, ensuring better efficiency for BECCS.

2.1 The worldwide waste structures

It is evident that waste treatment before B-V can be very complex and costly, but it can meet all requirements for environmentally friendly use and processing. The additional effort required is offset by the advantages associated with B-VM and there is no other system with this process and these capabilities.

The waste structure to be processed with its different fractions is crucial for the feasibility of using B-V worldwide. Global waste structures are similar when broken down to regionally determining compositions. This is particularly true after the locally necessary processing and even more so with the output from pyrolysis. In this context, the global waste composition and that from 17 representative cities in China, as well as in Germany, were closely examined.

Figure 2 [1] displays the structure of global waste. It is evident that at least about 3/4 consists of biomass and plastic; increasing to 4/5 if 50% of "Other" is covered by biomass and plastic. **Figure 3** [5] shows the municipal solid waste in 17 representative cities in China, displaying a very similar structure to **Figure 2** but with own water fraction (the whole German/Chinese study in [5]).

In Germany, the problem is that between 2000 and 2024, waste incineration has quadrupled, but biomass must be legally separated. This results in an annual amount

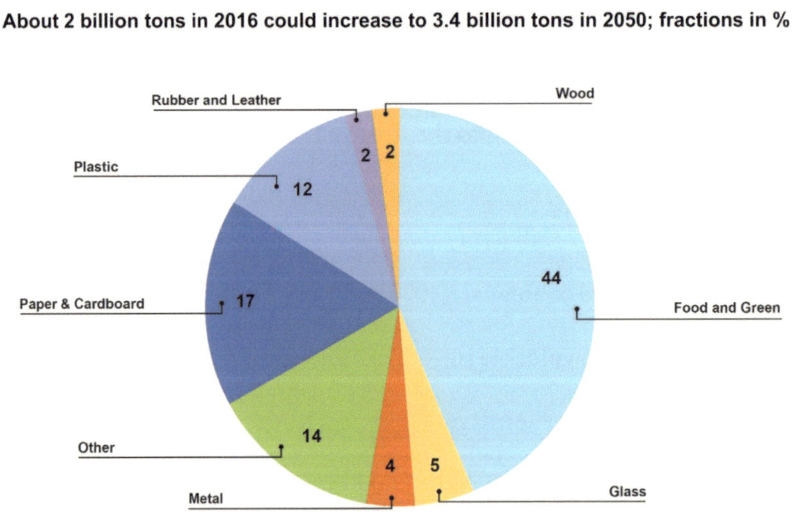

Figure 2.
Global waste with fractions. Based on figure 2.8 in [1]. This is an adaptation of an original work by The World Bank. Views and opinions expressed in the adaptation are the sole responsibility of the author of the adaptation and are not endorsed by The World Bank.

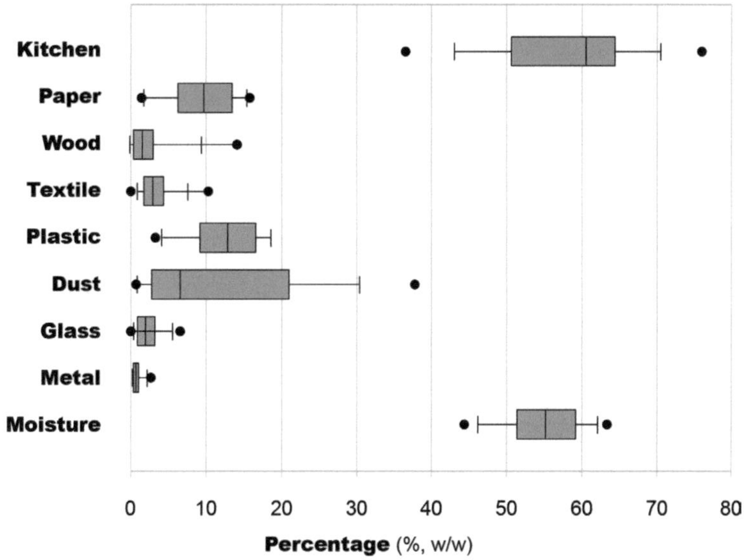

Figure 3.
Physical composition of MSW in typical cities of China (Yang Na, 2014; Yang et al., 2015): given as figure 5.5 in [5].

of about 16 million tons biomass, but besides and in addition, up to 40 % of household waste is still biomass and is therefore incinerated with household waste. The separated biomass is going to biomass processing plants to be burned, composted, or gasified. However, burning and gasification are inefficient and composting and gasification involve more or less contamination. The fact is that Germany and ultimately the EU cannot really handle the biomass they produce sustainably. It should be realized that at least all biomass collected from the households will inevitably always contain more or less foreign substances and also plastics. The data for Germany, regarding the composition of residual waste in 2002, show that biomass and plastics make up over 4/5 of the mass (**Figure 4** [6]) and this value is assumed to be realistic also for the present overall.

2.2 HERHOF's DRY STABILAT® as the basis for the pyrolysis input

Figure 4 illustrates the structure of the HERHOF DRY STABILAT® as a development that tries to modify the traditional incineration of waste. In Germany, there are extensive studies on waste processing and valorization through incineration, gasification and pyrolysis. This implies that some of the structures and pyrolysis data investigated there can be directly applied to the B-VM process, requiring minimal effort for additional proprietary and new research.

Currently, waste data from the HERHOF DRY STABILAT process (hereinafter referred to as HTS) are used in the B-VM concept. The HTS data are adapted as much as possible to the waste structures required for B-VM or can be converted. There are several scientific studies on HTS between approximately 1995 and 2015 (key references are [6–8]), including aspects related to pyrolysis. Most of the waste data required for B-VM are derived from these studies; this is particularly true for the pyrolysis input and the output, including data on heavy metals and pollutants. In the case of local waste, independent tests could be conducted for safety.

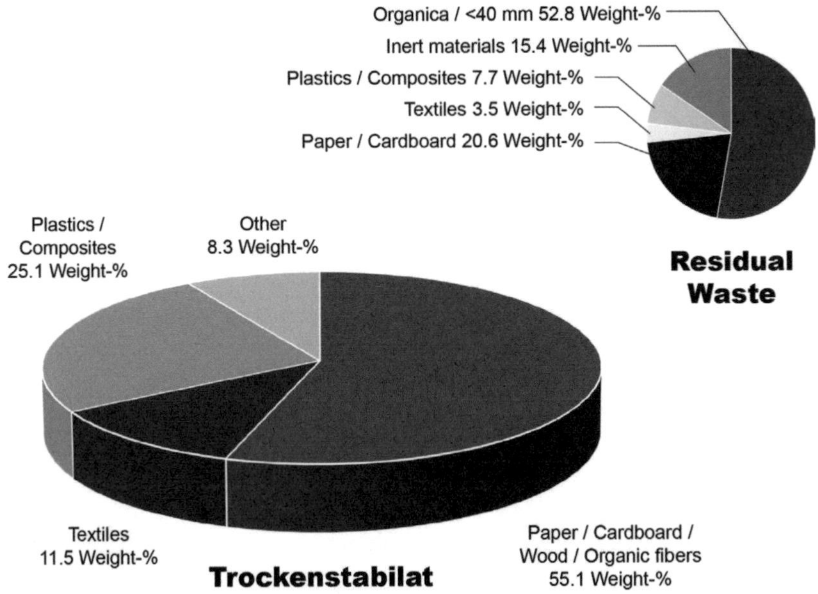

Figure 4.
Material composition of German residual waste processed to TROCKENSTABILAT®. Given in German as figure 7.4 in [6].

Contrary to widespread expectations in Germany, the waste recycling system developed by HERHOF proved to be practicable in the entire environment. However, the subsequent incineration and possibly the gasification of the produced STABILAT have not been convincingly implemented on a large scale. Importantly, systematic and EU-funded trials based on STABILAT with its low residual moisture content of ≤20% were conducted using a specially built pilot plant for several additional processing steps. Among other objectives, these trials aimed to produce clean biomass from STABILAT [9] and use it as environmentally friendly material. But the technical processing possibilities and their economic viability demonstrated in the "LIFE+ Project MARSS" (Material Advanced Recovery Sustainable Systems) are also crucial for the B-VM pyrolysis. It is essential that the demonstrated systems offer the new possibility to extract most of the inert material down to a few millimeters in size. If it is sensible to go further, it will be necessary to install specialized wet or air-based technologies.

2.3 Pyrolysis input data and output fractions

Based on data from China and Germany and with detailed further parameters for each fraction from **Figure 3**, initial general processing for the upstream system to B-VM was developed. Among other things a direct correlation between waste mass and engine power was established, indicating a requirement of about 10,000 t/a (wet and defined waste) per MW of engine power. For an actual plant, the locally given current waste structures must be considered. An example shows data from table in [8] with STABILAT material and proximate and ultimate analysis (both % weight and dry-base) including: moisture = 22.1; volatiles = 66.3; fixed carbon = 10.2; ash = 23.5; C = 42.1; H = 5.8; N = 1.3; S = 0.5; Cl = 1.1; O = 25.7 by difference; HHV = 18.42 kJ/kg, dry base). This structure is incorporated into the mass and energy balance we have developed for the B-VM and is shown in Section 4.4.

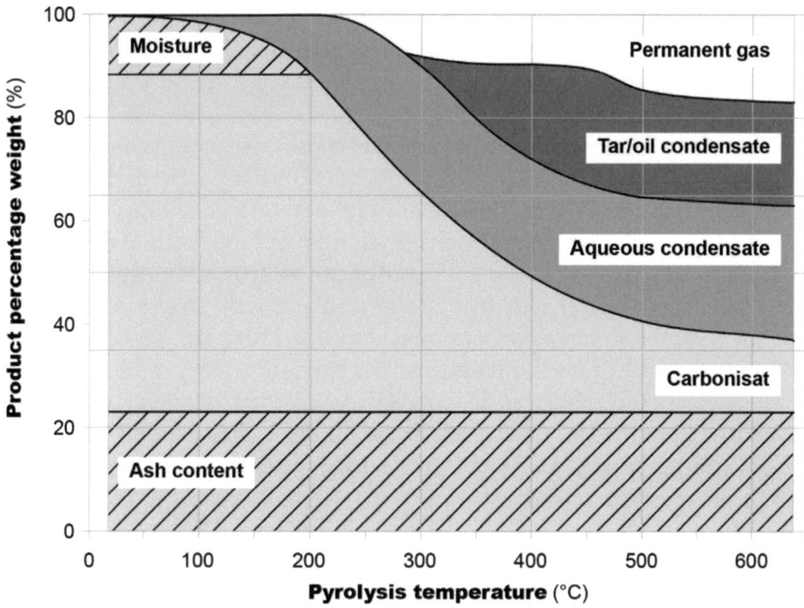

Figure 5.
STABILAT product fractions in the output of the rotary kiln with increasing pyrolysis temperature. Translated from figure 7.8 in [6] but without the scattered measurement results.

Figure 5 [6] displays the output fractions after the rotary kiln for STABILAT as weight percentages across rising temperatures (the B-VM operates in the range of approximately 450–600°C normally); the water content and the very high ash content of B-VM (including inert material) are important.

3. Decoupling thermal and electrical output

Based on **Figure 1**, the pyrolysis in the rotary kiln results in outputs of pyrolysis coke and pyrolysis gas (= "fluids"). The fuel for the large diesel engine comprises pyrolysis oil (on a new, large, commercial scale) and synthesis gas. The oil is derived by quenching from the pyrolysis gas and the synthesis gas is produced by gasifying the pyrolysis coke. Oil and coke thus serve as storable energy sources and can be used as a basis for motor fuel and/or for fluidized bed combustion (or similar) to generate thermal energy. With oversized storage units for coke and oil, it is possible to decouple the generation of thermal and electrical energy, achieving otherwise unattainable flexibility in the utilization of thermal energy in such motor systems.

The storage tanks for pyrolysis coke and oil are typically designed only in terms of capacity to compensate for predictable smaller fluctuations in the process before and after them inclusive of contracted normal change in power on demand for grid balancing. Expanding the storage tanks beyond this measure means that the entire system upstream and downstream of the storage tanks is more or less decoupled, with the interface lying downstream of the storage tanks. In extreme cases, the two areas thus created could also be physically separated and then connected through the transport of pyrolysis coke and oil. The upstream storage area must be designed for the expected delivery volume, fill the storage and be optimally utilized throughout the year. This also applies to the mass and energy ratio between pyrolysis coke and pyrolysis oil. The area downstream of the storage tanks must be designed in terms of the expected electrical and thermal energy to be generated in time, where it is important that the ratio between pyrolysis oil and synthesis gas as fuel for the diesel engine is very flexible.

The potential decoupling of thermal and electrical energy represents a completely new development and will have significant impacts. This is particularly true for the resulting capability of short-term and substantial network balancing, as shown in item 5. Similarly, the adaptation to the changing demand for thermal energy over longer periods means stored energy in the tanks can be accessed at any time as needed and by this even months later after production and also with large jumps in output.

One of the major issues in a conventional combined heat and power plant, for example, is the almost non-existent demand for thermal energy in the summer and the high-energy demand in winter. The necessary seasonal balancing can be easily achieved by adjusting the storage capacities for pyrolysis coke and oil. The upstream capacities for filling the storage tanks and the downstream capacities for combustion must then be determined accordingly.

4. Doubling net electrical efficiency of the power plant

The potential net electrical efficiency of the newly developed power plant is approximately doubled by employing a special large diesel engine instead of a steam turbine. Steam turbines used in waste incineration typically operate with an efficiency of ≤25%. In contrast, the diesel engine achieves an efficiency of nearly 55% and also has the best other parameters, besides other things boasting a guaranteed lifespan as the longest of any reciprocating combustion engine.

The comparison begins with a comparative presentation of parameters for waste incineration and the utilization of our B-V bioenergy process with upstream waste treatment, substantiating the data used as a foundation for combustion and our system. The basis for this is the actual net electrical efficiency achieved in waste incineration. Reliable empirical data are available for this value, which can also be calculated from existing data on waste and plant configuration. The efficiency of steam turbines is the determining factor for this value. The power plant's efficiency is only slightly lower due to minor losses through the aggregates used. In global waste incineration, capacity ranges are approximately per line with 20,000–380,000 t/a and about 4–40 MW per steam turbine. The relatively low turbine output is primarily due to the maximum possible steam temperatures and associated pressures resulting from the flue gases in waste incineration. The maximum turbine efficiency under these conditions is slightly above 30% (the highest known efficiency is realized in Amsterdam waste incineration plant with ≈ 30% net). However, with turbines specified in this way the German incineration plants for waste and for Refuse-Derived Fuel (RDF) only achieve net electrical efficiencies of about 11–15%. These for years 2012–2016 representative values were published by the German Federal Environment Agency (UBA) in 2018 [10] and can also be confirmed by calculations of 12–14% through corresponding mass and energy balances. It is important to note that the waste incinerated in Germany has a residual moisture content of about 25%. Globally, higher moisture content values reduce efficiency.

Compared to the steam turbine, the efficiency of the engine used in B-VM is the determining factor for the overall plant performance. The entire plant incurs greater efficiency losses due to its design and the systems and units used. The actual achievable net electrical efficiency can only be more accurately determined in advance through mass and energy balances, including all parts of the process.

4.1 Drive unit by slow-running, two-stroke diesel engine

The drive unit utilizing B-VM necessitates a much higher efficiency than that of a steam turbine. The following **Figure 6** [11] illustrates the application of some of the most commonly used power generation technologies along with their respective performance ranges in megawatts (MW) and the achievable efficiencies. Although this diagram of 2016 is relatively up-to-date and accurate, it is missing, as in nearly all similar figures and the associated discussions and evaluations, the inclusion of the slow-running, two-stroke large diesel engine, which is produced with generator outputs ranging from 3 to 80 MW and achieves efficiencies of up to approximately 55% (see **Figure 7** [12]). **Figure 6** alone by its additions underscores the crucial objective of replacing the turbine with two-stroke diesel engine.

The slow-running, two-stroke large diesel engines are evidently capable of surpassing the current application range of steam turbines for waste combustion in every respect and by a considerable margin. These large diesel engines operate with undisputedly the highest availability, efficiency, reliability, and lifespan (guaranteed up to 25 years at 8000 hours per year). These have been deployed for decades in the largest ships and in stationary power plants, especially in island operations and achieve the highest fuel efficiency in propulsion systems. Only the use of pyrolysis oil on a large commercial scale would represent an innovation. Initial tests for this were conducted in 1994 and have been confirmed to this day by the two engine developers who have made respective proposals. Due to the severe consequential damage in the event of a failure or extended repair times, the engines were designed to achieve the highest MTBF (Mean Time Between Failure = "Reliability") and the lowest MTTR (Mean Time To Repair; where then the following applies:

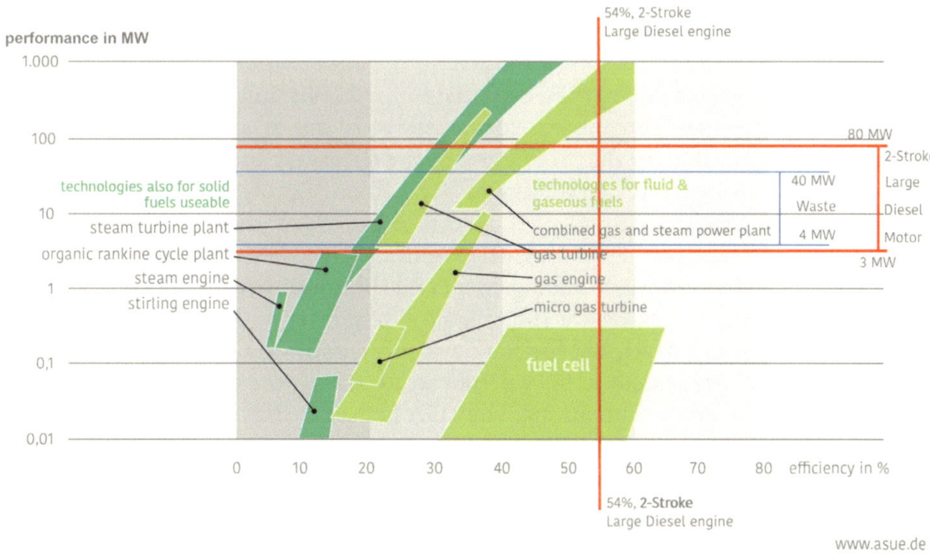

Figure 6.
Technologies for generating electricity with their power and efficiency ranges. Basis given in German by ASUE in [11].

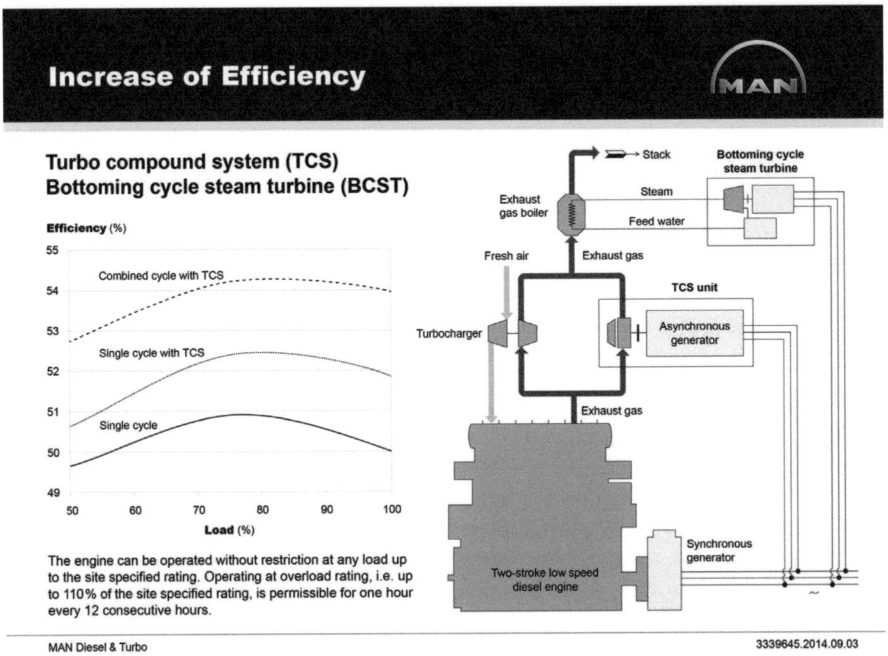

Figure 7.
Increase of efficiency, MAN low-speed, two-stroke diesel engine with turbo compound system (TCS) in [12].

"Availability" = MTBF/(MTBF + MTTR)) and this even at peak continuous output. At the same time, for economic reasons, a long lifespan (for optimal ownership costs) and low fuel consumption are required. This would also lead to environmental benefits over decades.

4.2 Pyrolysis by rotary kiln

The externally heated rotary kiln, essential for pyrolysis, in conjunction with the two-stroke large diesel engine, constitutes the most important unit in the B-VM system and is responsible for the production of the necessary fuel. A highly comprehensive study published at the beginning of 2022, titled "Pyrolysis-based Municipal Waste in Poland - SWOT Analysis" [13], asserts that... "Pyrolysis is an economically attractive alternative to incineration, with a significantly lower environmental impact, allowing efficient waste management and the use of pyrolysis by-products ...."

In Germany, there are extensive and reliable experiences with some industrial waste pyrolysis plants (MPA), but further developments and utilizations are lacking. Nonetheless, these systems clearly demonstrate at least the feasibility of rotary kilns in waste management operations. The MPA in Burgau/Bavaria was operational from around 1985–2015, handling up to 35,000 tons each year and serving the waste disposal needs of approximately 120,000 residents and industries, including sewage sludge. The pyrolysis coke was landfilled and the landfill gas produced was combusted along with the pyrolysis gas from the rotary kilns in a combustion chamber at approximately 1200°C to generate steam. Two rotary kilns, each with a diameter of 2.2 m and a heated length of 21 m, were employed. The CONTHERM MPA was used as an upstream unit for a coal-fired power plant for about 10 years. The two rotary kilns, 20 m in length and 2.8 m in diameter, processed up to 100,000 tons each year of RDF (Refuse-Derived Fuel).

Besides the influence on the distribution of the output fractions (**Figure 5** [6]), the pyrolysis temperature also has an effect on the pH value of the pyrolysis oil. The residence time of the input and the formed pyrolysis gases in the rotary kiln, as well as the heating rate, play significant roles and can be manipulated to control the output from the rotary kiln. Produced pyrolysis coke, in combination with hot pyrolysis gas and appropriate process control through catalytic effects, can ultimately lead to higher-value output in pyrolysis oil and provide an advantage for engines (see "TCR" process with its thermo-catalytic reforming [14]). The "Co-pyrolysis of Biomass and Plastics from Municipal Wastes" (search these keywords) generally yields a more advantageous pyrolysis output than individual fractions and there is a vast array of procedures, scientific experiments and various material combinations available. The suitability of different waste components for pyrolytic conversion has been demonstrated through small-scale experiments, which also showed that the varied composition of waste does not play a decisive role and can generally be compared well with international data [15].

4.3 Further crucial units

Most important are quenching of pyrolysis gas, gasification of pyrolysis coke and handling of thermal energy. The methods for implementing these can be covered with a large number of well-known variants and decades of experience. All technologies used in B-V and B-VM are designed for comparatively safe and long-term use align with the current state of the art and largely comply with the use of sustainable input and operational materials in the process.

In the quenching zone, the hot gases produced by pyrolysis are rapidly cooled, halting all ongoing reactions and directly converting the gases into a permanent gas fraction and a pyrolysis oil fraction. When the hot gases enter the quencher, the pyrolysis gas still contains a considerable amount of dust and water, thus requiring appropriate treatment. The permanent gas serves as a heating base for the rotary kiln. The produced pyrolysis oil depends among other things on the conditions during quenching and may need further processing for buffering in the tank system and subsequent use in the engine (similar to the use of heavy oil or ORIMULSION for this engine type).

For the gasification of pyrolysis coke into the required synthesis gas, we generally use fluidized bed gasifiers. It is crucial that gasification of coke, theoretically devoid of volatile components, produces almost tar-free synthesis gas. All our previous approaches have been based on Auto Thermal fluidized bed gasification, where the necessary thermal energy is generated by burning part of the coke within the system. Water and almost pure oxygen are used as gasifying agents. In the future, the use of a Dual Fluidized Bed (DFB) will be explored. This will be the case in particular if the fundamental possibility of using gas engines instead of diesel engines is utilized. Among other things, the combustion of part of the coke and the use of oxygen will be eliminated and the entire process could probably be made more efficient and effective. But on the other hand, gas engines do not achieve the high performance and efficiency of the two-stroke diesel engine (see **Figures 6** and 7 [11, 12]). In this context, the entire mass and energy balance of B-VM needs to be redesigned and the use of a quencher also reviewed.

For thermal handling, a significant area arises from the mass and energy flows, linked with the blue flows as shown in **Figure 1**. In this depiction, all areas of thermal energy have been simplified for clarity. The aim is to comprehensively capture and optimally utilize individual items of generated and required thermal energy. The underlying mass and energy balances are complex and extensive, ultimately utilizing thermal oil. The waste heat from the heating system at the rotary kiln can be considered as a base. This is primarily combined with waste heat from engine exhausts and the heat of the fluidized bed exchanger. Merged, this thermal energy is then directed either to internal or external consumers or used in a separate power generation engine, particularly in the form of an ORC turbine or a steam engine, both of which exhibit excellent performance in the required partial load range.

4.4 Resulting mass and energy balance MEB

A detailed MEB for the process shown in **Figure 1** can be used to derive CAPEX and OPEX and demonstrate a self-contained and resilient system. The feasible profitability calculation allows the creation of the necessary business plan and with the appropriate engineering a pilot plant can be designed.

Based on the B-V, a business plan for biomass and a 14 MW engine was developed, implementing the Renewable Energy Sources Act (EEG) in Germany. The net electrical efficiency was thus represented with the Mass and Energy Balance (MEB) at 36%. Approximately 600 items of the MEB were evaluated and results in a CAPEX of about 100 million Euros (excluding BECCS). The Return on Investment (ROI) is 11% and the Internal Rate of Return (IRR) reaches nearly 20%. The entire concept and calculation have a "Study Estimate Level" with an accuracy of +/−20% to +/−30%. On this basis, a similar B-VM concept for Poland and for the Czech Republic determines a net electrical efficiency of about 30% and a CAPEX of about 120 million Euros (including BECCS; approximately 3.8% of the total net electrical efficiency of 33.8% is allocated for the upstream waste processing). **Figure 8** shows an excerpt from this structure of the MEB (excluding BECCS and excluding waste processing). The waste fees and revenues from all extracts (including CO_2) then lead to significantly improved economic conditions compared to the B-V.

Due to the stipulations of the German EEG, it is not possible to directly transfer the data to other countries. However, CAPEX and OPEX can be adjusted country-specifically and then work with the respective location-specific fees for waste and the compensation for electricity and heat (plus the revenues from the extracts, including CO_2). The decision to develop the B-VM derived from the B-V was, among other things, based on an analysis of the structure and CAPEX of waste incineration. It was found that only cost-effective providers from China could compete regarding the CAPEX of the B-VM.

In addition to clear ecological advantages over waste incineration, the significantly higher net electrical efficiency and the possible thermal decoupling lead to a strong global unique selling proposition (USP). The possible use of BECCS increases this advantage. Investors and plant operators benefit from the previously unattainable performance and environmental friendliness compared to waste incineration.

5. Possibility to realize negative emissions through BECCS

The utilization of BECCS is facilitated by the high biomass content in waste and the high efficiency of electricity generation, as well as the decoupling from thermal energy.

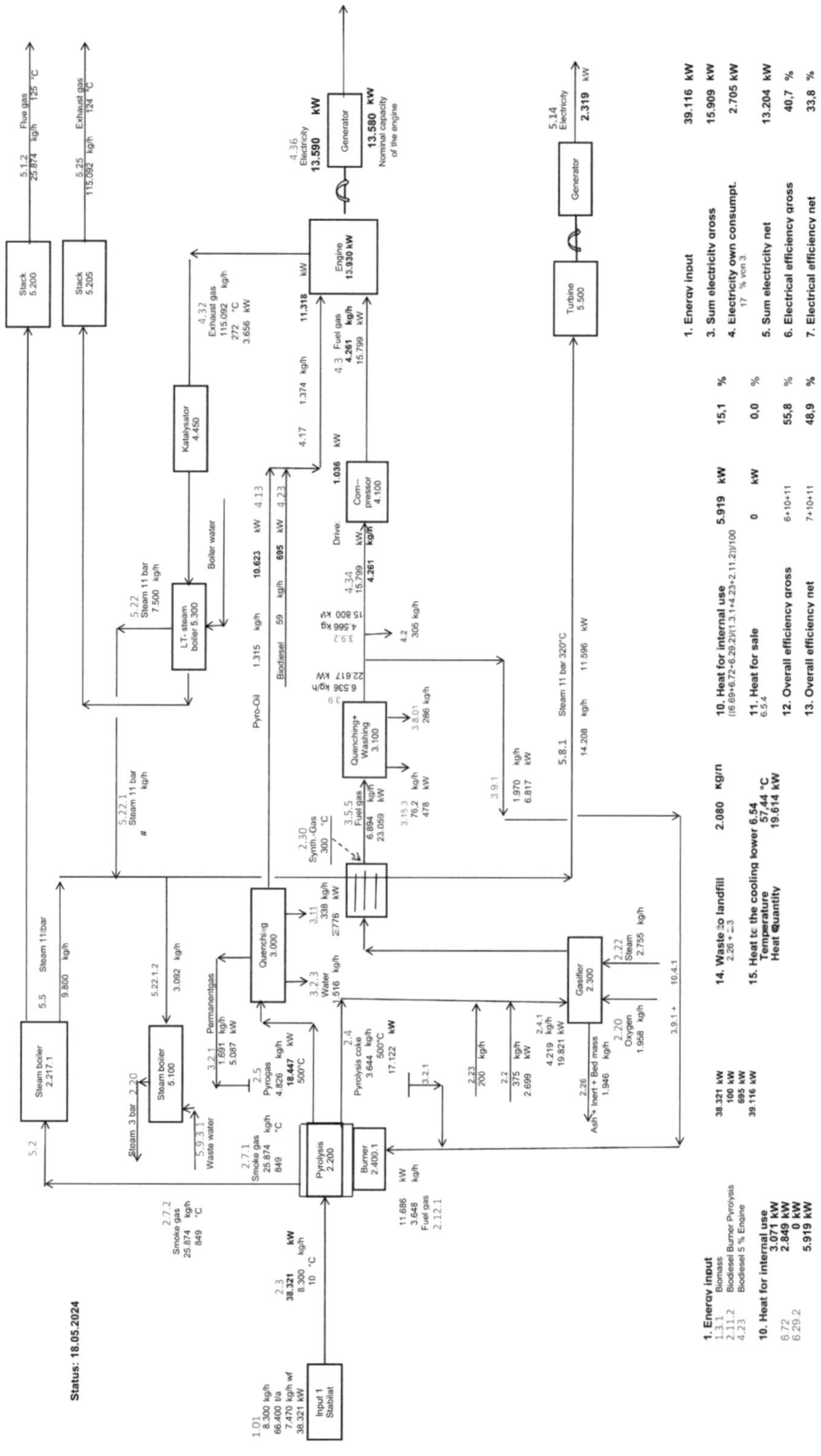

Figure 8.
Simplified mass and energy balance for STABILAT and slow running, two-stroke, dual-fuel 14 MW diesel engine. From author's BMC Sp.z o.o. Working Group B-VM.

Additionally, locally available biomass can be used in the decentralized facilities. Biomass in the B-VM is climate-neutral and the use of BECCS can enable the achievement of negative CO_2 emissions. This also allows for a balance with the CO_2 from the plastic content in the waste, thereby realizing more or less negative CO_2 emissions. BECCS with the B-VM presents an opportunity to implement Negative Emission Technologies (NETs) for CO_2 on a large scale. There are significant objections to BECCS globally due to its associated consumption of agricultural land, water, and fertilizer etc. However, these do not apply to the B-VM for our technology is scalable and the required biomass is based on waste.

Preliminary calculations suggest that up to 1 billion tons of CO_2 per year could be extracted from the exhaust gases of a 50% portion of the globally possible B-VMs. This corresponds to about 20% of the 5 billion tons per year of negative emissions postulated by the IPCC (Intergovernmental Panel on Climate Change) and NAS (National Academy of Sciences) for BECCS.

Further rough calculations indicate that the additional revenue from power generation could at least cover the expenses for CO_2 extraction. The effort for further utilization of CO_2 then depends on the respective local conditions. Economically the B-VM process, with its very high net electrical efficiency and variable thermal energy and the expected CO_2 compensations, advocates because using BECCS in the B-VM as the best complementary investment to enable better amortization.

The output of CO_2 by the HTS as STABILAT with 10% water and a 14 MW diesel engine was roughly calculated and shows:

The collection of a waste input of about 133,000 t/a results in 66,400 t/a entering the pyrolysis process and water-free per hour: C = 3145 kg; O = 1920 kg; H = 433 kg; N = 97 kg; Cl = 82 kg, S = 37 kg; ash = 1755 kg.

1. Flue gas with 26 t/h contains 3.6 t/h CO_2, at 125°C

2. Engine exhaust gas with 115 t/h contains 3.5 t/h CO_2, at 124°C

With 8000 operating hours per year, this amounts to approximately 57,000 tons of CO_2, of which at least could be extracted an estimated 50,000 t/a. If these figures are linearly scaled up to the largest diesel engine of 80 MW, a corresponding facility would emit about 326,000 t/a CO_2, of which could be extracted annually nearly 290,000 tons.

The extraction of CO_2 from exhaust gases is already known for a long time and the development of more efficient processes is ongoing worldwide. For the B-VM, we consider working with the Chilled Ammonia Process (CAP) and are monitoring developments such as the Metal-Organic Framework (MOF). Local conditions are always crucial for the realization of storage or utilization of CO_2 and several options are currently available and new ones under development.

6. Base and peak load and very strong energy balancing

The B-VM facilities are decentralized in their establishment and are broadly distributed according to the given concentration of population and economic activity. This configuration evidently allows these plants, due to their complete independence from external influences, to serve as ideal compensators for regional grid fluctuations and also to absorb extreme outages, for example, from wind power and, above all, photovoltaic sources.

Based on the size of the buffer storage tanks, the system can cover peak performances and thus be utilized in the normal area of "balancing energy" to compensate for electricity network fluctuations. This is applicable both for the use of the large diesel engine and for the steam-powered units. Particularly, an Organic Rankine Cycle (ORC) system is capable of responding very quickly and with its full capacity. In the case of the large diesel engine, a significant portion of its output can be utilized with minimal efficiency losses according to its characteristics curve (**Figure** 7 [12]), though with time constraints in the overload range.

If necessary, this capability can be enhanced by a significant enlargement of the buffers for pyrolysis coke and pyrolysis oil and consequently serve to split the overall facility into two systems. Including the two buffers, the plant can be designed for optimal use with its annual input of 8000 hours, thereby continuously filling both buffers. Concurrently, however, based on the two buffers, a system for extraction for the delivery of electricity and thermal energy is established, whereby the annual usage time is more or less reduced. In return, the system performance is increased. For instance, with an operational time of 4000 hours per year, it would be possible to double the engine power during the night (using a 28 MW engine instead of a 14 MW engine) and thus provide an optimal compensation example for the loss of photovoltaic power at night.

Assuming a waste capacity of 25 million tons in Germany and other regionally available biomass of the same amount, it would be feasible to provide about 50% of the power currently generated by photovoltaics during the day also at night with approximately 410 systems and 28 MW engines.

7. Conclusion

The importance of the work lies above all in showing that there is a safe thermal process for recycling waste which is far superior to incineration in ecological and economic terms and which can then be used to take real and successful action against the worldwide increase in waste. It is worth noting that this process enables the global use of BECCS for the first time due to its decentralized high proportion of biomass in the waste and the best electrical efficiency and the possibility of decoupling from thermal energy. This is particularly the case if the respective location offers the possibility of simply processing or storing the extracted CO_2.

The B-VM power plant has the potential to solve the existing problems associated with waste and its incineration and also offers new and best opportunities. This applies not only to the approximately 20% of waste incinerated worldwide today, but to almost the whole waste worldwide. Notably, the predominant current handling of biomass could become an obsolete method, facilitating the elimination of composting which invariably releases pollutants into the environment to varying degrees. Additionally, the incineration of materials such as plastics, in cement kilns, with its associated ecological problems, becomes unnecessary. Furthermore, the environmentally very harmful open production of methane from the fermentation of free biomass would at least be reduced. With **Figure 1**, there also exists a fundamental opportunity to utilize the intermediate product, synthesis gas, for producing alternative fuels such as DME (Dimethyl Ether) and OME (Oxymethylene Dimethyl Ether) or similar through appropriate processes.

The B-VM process offers the possibility to utilize existing systems and units with many years of successful use and an assessment as state-of-the-art technology; only their configuration and the use of pyrolysis oil are new. This means besides other things that all the necessary data is known and can be found in existing systems. Precise and detailed explanations and verifications exist for the process and its evaluation based on the planned project for B-V according to the Renewable Energy Sources Act (EEG) in Germany and the derived B-VM projects in Poland and the Czech Republic. The prototype was designed as fictional large-scale pilot plant with a comprehensive and detailed mass and energy balance. Therefore, the remaining tasks include developing a location, adjusting all the capacities, conducting detailed engineering and realizing a first pilot power plant with profitability calculation.

The opportunity to utilize BECCS arises from the high proportion of biomass in waste and the very high efficiency of electricity generation, along with the straightforward decoupling from thermal energy. All the disadvantages typically associated with BECCS up to date, such as significant loss of agricultural land and consumption of fertilizers and water, as well as logistics costs, is avoided.

Given the potential for rapid and consistent implementation of BECCS, the usual governmental responsibility associated with waste proves advantageous. Since BECCS can be assessed as an extension investment in waste recovery, the current problems of waste disposal and the achievement of CO_2 reductions can be linked and implemented by the state.

In this regard, it would be sensible to employ large and global capacities in engineering and production for realizing the B-VM without delay and in this aspect some companies have a great and unique alternative opportunity with the growing elimination of fossil fuel projects.

Author details

Hubert Boeckmann†
BMC Sp.z o.o. Working Group B-VM (= Bioenergy Process B-V with Upstream Waste Processing), Jezow Sudecki, Poland

*Address all correspondence to: rittergut-deensen@t-online.de

† Data based to this study are available on request from the author.

References

[1] Silpa K, Yao L, Bhada-Tata P, Van Woerden F. What a Waste 2.0: A Global Snapshot of Solid Waste Management to 2050. Urban Development Series. Washington, DC: World Bank; 2018. DOI: 10.1596/978-1-4648-1329-0

[2] Schneider A, Cehreli E. Verfahren und Vorrichtung zur Produktion von Synthesegas und zum Betreiben eines Verbrennungsmotors damit (WO 2011/110138 A1). EPA; 2011

[3] Schneider A, Cehreli E. Method and device for producing synthesis gas and for operating an internal combustion engine therewith (US 9,057,033 B2). U.S. Patent and Trademark Office. 2015

[4] Çehreli E, Schneider A. Konzepte für die nachhaltige und CO2-neutrale Energieerzeugung aus organischen Rest- und Abfallstoffen sowie NONFOOD-NawaRo. In: Maus W, editor. Zukünftige Kraftstoffe, ATZ/MTZ. Berlin, Heidelberg: Springer Vieweg; 2019. DOI: 10.1007/978-3-662-58006-6_14

[5] Nelles M. Recycling and Recovery of the Biogenic Fractions from Municipal Solid Waste in the PR of China. With Physical Composition of MSW in Typical Cities of China from Yang Na, 2014; Yang et al., 2015. Rostock: Studie im Eigenverlag der Uni. Agrar- und Umweltwissenschaftliche Fakultät in Rostock; 2017

[6] Niederdränk J. Untersuchung zur Gewinnung von Carbonisaten aus der thermischen Veredlung mechanisch-biologisch stabilisierter Restabfälle. In: Schriftenreihe zur Aufbereitung und Veredelung 7. Aachen, RWTH. Aachen, Germany: Shaker Verlag; 2002

[7] Gummersbach J. Experimentelle Untersuchung der Pyrolyse in einem Drehrohr als Bestandteil einer Vergasungsanlage mit Restkoksverbrennung. In: Berichte aus der Energietechnik. Aachen, Germany: Universität Siegen, Shaker Verlag; 2005

[8] Dunnu G, Panopoulos KD, Karellas S, Maier J, Touliou S, Koufodimos G, et al. The solid recovered fuel Stabilat®: Characteristics and fluidised bed gasification tests. Fuel. 2012;**93**:273-283. DOI: 10.1016/j.fuel.2011.08.061

[9] Giani HM. Biomassebrennstoff aus Hausmüll Ergebnisse des EU-LIFE+ Vorhabens MARSS. Müll und Abfall. 2015;(12):660-663. DOI: 10.37307/j.1863-9763.2015.12.05

[10] Umweltbundesamt UBA. Energieerzeugung aus Abfällen Stand und Potenziale in Deutschland bis 2030. Editor UBA in Dessau-Roßlau; 2018. 51/2018, 159

[11] ASUE. Efficiency Comparison in Power Generation—Graphic from the editor ASUE in Berlin, brochure "CO_2 avoidance" 2016. Translation into English and all changes and reproduction with consent of ASUE

[12] MAN. Presentation MAN B&W. Copenhagen: With internal paper MAN Energy Solutions; 2016

[13] Ławi'nska O, Korombel A, Zajemska M. Pyrolysis-based municipal solid waste management in Poland—SWOT analysis. Energies. 2022;**15**:510. DOI: 10.3390/en15020510

[14] Elmously M, Jäger N, Neidel J, Apfelbacher A, Hornung A. Upscaling of thermo-catalytic reforming process

from lab to pilot scale. Industrial and Engineering Chemistry Research. 2019;**58**:15853-15862. DOI: 10.1021/acs.iecr.9b00765

[15] Bradfield F. Examination of the Thermal Properties of Municipal Solid Waste and the Scalability of its Pyrolysis. Master from Stellenbosch University, in Stellenbosch. 2014. Available from: http://scholar.sun.ac.za

Chapter 2

Navigating the Global Challenge through Comprehensive Analysis of Industrial Effluents, Environmental Impacts, and Sustainable Solutions

Murugesan Manikkampatti Palanisamy, Padmapriya Veerappan, Kalaivani Ramanathan, Sankaranarayanan Devikarani, Dharunya Rajagopal, Kokila Kannusamy, Saranya Indrabal and Anusha Kolanji

Abstract

In contemporary times, the treatment and disposal of industrial waste present formidable challenges on a global scale. This study meticulously investigates the complexities surrounding industrial effluents, addressing the diverse range of compositions and their far-reaching consequences on ecosystems across various industries. Employing an exhaustive analysis of existing literature, the chapter elucidates the intricate mechanisms through which industrial activities contribute to environmental pollution, encompassing both immediate and long-term effects within all industry sectors. Additionally, the review scrutinizes prevailing regulatory frameworks and cutting-edge technological advancements designed to alleviate the adverse impacts of industrial effluents. The chapter aims to synthesize information from a myriad of sources, providing invaluable insights into the multifaceted challenges posed by diverse industrial effluents. Beyond identification, the study lays the groundwork for the development of sustainable solutions, emphasizing a detailed analysis of various treatment technologies. Furthermore, the chapter offers insightful suggestions to stimulate research activities among scholars, contributing to the collective pursuit of effective strategies for environmental protection and sustainable industrial practices.

Keywords: industrial effluents, global analysis, worldwide challenges, environmental impacts, treatment technologies, regulatory frameworks

1. Introduction

Effluents discharged from various industries pose significant environmental challenges, particularly in terms of water pollution and resource degradation.

Effective treatment of these effluents is crucial to ensure sustainable industrial operations and minimize environmental impacts. This report provides a graphical analysis of effluent treatment efficiencies across several industries to assess their contribution toward environmental stability and highlight areas for improvement. This chapter presents an overview of current wastewater treatment methods applied across various industries, highlighting their effectiveness in pollutant removal and proposing future advancements. The research focuses on diverse effluents, including dairy, beverage, sugar, chemical, confectionery, and textile industries, among others. Conventional treatment methods such as aerobic and anaerobic processes, chemical oxidation, and advanced filtration techniques are assessed for their removal efficiencies. Despite varying success rates, with some industries achieving up to 100% removal, the need for innovative technologies like membrane systems, sequencing batch reactors, and sustainable methods such as phytoremediation is emphasized to enhance treatment efficacy and environmental sustainability.

2. Impact of industrial effluents on the environment, human beings, animals, and other living organisms

Industrialization is a critical factor in the development of any country. Industrial estates, established to meet the increasing demands of growing populations, contribute significantly to the economy. However, this development often comes at a high environmental cost. The introduction of industries, while manufacturing useful products, simultaneously generates waste in solid, liquid, and gaseous forms, which leads to environmental hazards, Soil, Air, and Water pollution, and energy losses [1]. Over the last several decades, rapid industrial growth has disrupted the environment, causing profound ecological imbalances.

Electrical power plants primarily burn fossil fuels to generate electricity. During the combustion process, hazardous gases and particulate matter are released, contributing to air pollution. The waste heat generated by the plants also has detrimental effects on nearby water bodies, disrupting aquatic ecosystems. Furthermore, ash and other residues from the burning process are potential sources of soil and water contamination [2]. In addition to that, battery manufacturing industry is considered one of the most toxic and hazardous industries. The production of batteries involves various toxic chemicals such as lead, cadmium, and sulfuric acid, which pose significant risks to the environment and human health. Improper disposal of battery effluents contaminates soil and groundwater, causing long-term environmental degradation [3]. Nuclear power plants generate radioactive effluents that are released into the environment in both airborne and liquid forms. The radioactive substances from nuclear fission, such as tritium and iodine-131, can contaminate air and water, leading to severe consequences for human health and biodiversity. The long-lasting nature of radioactive materials further exacerbates environmental pollution [4]. Reported waste from mines and quarries is highly toxic and contaminates natural water resources. Toxic metals such as arsenic, lead, and mercury leach into water bodies, posing a threat to aquatic ecosystems and human health. Long-term exposure to these contaminants can cause chronic diseases, including cancer and neurological disorders [5]. Hence, petroleum industry produces both gaseous and liquid pollutants. While gaseous emissions like sulfur dioxide contribute to air pollution, liquid effluents can pollute water bodies. Although Effluent Treatment Plants (ETP) can treat these liquid pollutants, incomplete or inefficient treatment processes still allow harmful substances to enter the environment, affecting both human and animal populations [6]. Chemical industries release wastewater laden with high concentrations of organic and inorganic pollutants, some of which are toxic to both humans and animals. These pollutants include volatile organic compounds (VOCs), which can cause respiratory problems and contribute to smog formation, severely affecting air quality and public health. The wastewater contains toxic chemicals such as phenol, cyanide, and ammonia, which are harmful to aquatic ecosystems. These effluents reduce dissolved oxygen (DO) levels in water bodies, killing aquatic life and damaging ecosystems. This was discussed in **Table 1** for various effects related to the environment and others.

The food industry generates wastewater with high levels of Chemical Oxygen Demand (COD) and Biological Oxygen Demand (BOD), which can deplete oxygen in receiving water bodies, causing harm to aquatic life. The large volume of water required in food processing exacerbates the environmental impact [7]. Also reported, Wastewater from the dairy industry is rich in organic matter, mainly from milk residues, and can lead to the eutrophication of water bodies. This can result in algal

Sl. No	Industry effluent	Effects	Reference
1	Electrical power plant	Fuels are transported, pulverized, and stored to enhance efficiency. Combustion generates heat for steam production.	[2]
2	Battery Manufacturing	The battery industry is highly toxic and hazardous. Batteries consist of positive and negative electrodes and an electrolyte. The production process includes oxide milling, grid casting, paste preparation, curing, and assembly.	[3]
3	Nuclear Power Plant	Nuclear plants release radioactive effluents in both airborne and liquid forms, primarily through nuclear fission. Both Pressurized Water Reactors (PWR) and Boiler Water Reactors (BWR) emit these pollutants.	[4]
4	Mines and Quarries	This industry contaminates natural resources, particularly water. The wastewater from mines and quarries contains highly toxic metals, leading to diseases in affected areas.	[5]
5	Petroleum Industry and petrochemical	These industries produce gaseous and liquid pollutants. Gaseous emissions are more manageable, while liquid pollutants require treatment in Effluent Treatment Plants (ETP) to prevent environmental contamination.	[6]
6	Food Industry	The food industry produces wastewater with high levels of COD and BOD, contributing to water pollution. This industry also consumes large quantities of water throughout its processes.	[7]
7	Dairy Industry	Dairy wastewater contains significant amounts of organic solids, including casein and salts from milk, which can negatively affect the environment.	[8]
8	Oil extractive mill	These mills generate wastewater primarily during the bleaching process. The waste often needs to be treated in Effluent Treatment Plants (ETP) to reduce its environmental impact.	[9]

Sl. No	Industry effluent	Effects	Reference
9	Leather Industry	The leather tanning process releases chemicals that convert raw materials into durable leather, but also creates waste that can harm the environment.	[10]
10	Pharmaceutical Industry	Wastewater from pharmaceutical production contains toxic chemicals harmful to humans and animals. Various treatment methods, such as membrane filtration and reverse osmosis, are used to reduce pollution.	[11]
11	Agriculture Industry	The agricultural industry uses large volumes of water and produces wastewater that contains organic matter, dissolved minerals, nutrients, chemicals, and pathogens, all of which pose environmental risks.	[12]
12	Textile Industry	Effluent treatment in the textile sector involves removing suspended solids and treating wastewater through screening, sedimentation, and chemical coagulation to reduce pollution levels.	[13]
13	Paper and Pulp Industry	This industry uses a combination of primary, secondary, and tertiary treatments, including sedimentation, flotation, biological treatment, and membrane filtration, to manage wastewater.	[14]

Table 1.
Summary of Industrial Effluents and Their Environmental effects".

blooms, which reduce oxygen levels and kill aquatic organisms [8]. A huge rate of wastewater generated from oil extractive mills, particularly during the bleaching process, contains high levels of organic pollutants. Improper treatment and disposal of this wastewater can contaminate local water supplies and harm aquatic ecosystems [9]. Agriculture industry generates large volumes of wastewater containing organic matter, nutrients (e.g., nitrogen and phosphorus), and pesticides. These pollutants can contaminate water bodies, leading to eutrophication and harm to aquatic life. Additionally, the use of chemical fertilizers and pesticides can result in soil degradation and loss of biodiversity [12].

The tanning process in leather industries releases large amounts of chemicals, including chromium and sulfides, which are toxic to both the environment and living organisms. Improper disposal of these effluents can contaminate soil and water resources, leading to serious health issues for humans and animals [10]. In previous studies reported, the pharmaceutical industry discharges wastewater containing toxic chemicals, antibiotics, and hormones, posing significant threats to both human and animal health. Treatment processes such as membrane filtration, reverse osmosis, and coagulation can reduce the concentration of harmful substances, but improper treatment still poses significant risks [11]. Textile industry effluents contain dyes, heavy metals, and suspended solids that can contaminate water bodies. Treatment processes such as sedimentation, flocculation, and chemical coagulation are used to manage these effluents, but inefficient treatment can result in environmental pollution [13]. Paper and pulp industry generates large quantities of wastewater that contain organic matter and chemicals used in the pulping process. Advanced treatment methods, including biological treatment and membrane separation, are employed to manage these effluents. However, improper handling can result in significant environmental degradation [14]. Industrial effluents have a significant impact on the environment, human health, and biodiversity. The contaminants released from various industries lead to air, water, and soil pollution, posing long-term risks to ecosystems. Effective wastewater treatment and regulatory frameworks are necessary to mitigate the adverse effects of industrialization.

3. Worldwide challenges for industrial effluent treatment

Recycling spent batteries plays a crucial role in resource conservation and environmental protection, but it faces challenges such as limited recycling channels, high costs, and technical complexities. To tackle these challenges, a review has been conducted to highlight the significance and potential value of battery recycling [15]. However, battery waste management poses significant challenges, primarily because inflow rates of waste batteries are not well-classified by the industry, leading to many batteries ending up in landfills. This improper disposal has adverse economic, health, and environmental impacts [16]. Key challenges in battery waste disposal include the collection, safe transport, and storage of used batteries [17]. Guidelines are necessary to ensure the safe discharge of active batteries before transport to collection points, as partially charged batteries pose fire and short-circuit risks. Immersing spent batteries in an aqueous solution can mitigate these risks by reducing battery temperature during discharge [18]. Effective battery segregation by chemical composition is essential during processing [19]. Impurities complicate the metal recovery phase, necessitating data sheets from manufacturers for each recovered chemical to facilitate segregation [17]. Safe dismantling of batteries is another concern, especially in unorganized sectors where workers often handle batteries without proper safety measures. These workers typically recover casings, copper, aluminum foil, and circuit boards while treating cathode and anode black powder as waste, which is improperly disposed of in landfills or water bodies [20].

Most process routes successfully achieve high yields for valuable metals like cobalt, copper, and nickel. In contrast, lithium, despite its significant economic value, is recovered in only a few processes and with lower efficiency. Although recovering low-value components such as graphite, manganese, and electrolyte solvents is technically possible, it presents economic challenges. Additionally, managing organic and halogenic components across all process routes leads to technical complexities and elevated costs [21]. Developing environmentally friendly technology for recycling, recovering, and reusing battery metals is crucial. Conventional hydrometallurgy, which uses harsh acidic or basic conditions, provides good leaching and recovery rates but generates hazardous fumes and requires substantial water and neutralizing solutions, increasing wastewater treatment costs and resource wastage [22]. Below **Table 2** is a discussion of Various Industrial Effluents Treatment methods/techniques.

There is a need for the waste management industry to classify the inflow rates of waste batteries accurately, aiding in better planning and management of battery waste and reducing the number of batteries ending up in landfills [16]. Establishing comprehensive guidelines for the safe discharge of active batteries is crucial; proper protocols should be in place to prevent fires or short circuits during transport, and techniques such as immersing spent batteries in an aqueous solution should be standardized to mitigate risks [18]. Implementing stringent segregation processes based on chemical composition is necessary to facilitate efficient recycling, with manufacturers providing detailed data sheets for each chemical used to aid in accurate segregation [17, 19]. There is an urgent need to improve safety

S.No	Industry Effluents	Treatment Method Followed	Removal Rate %	Future Recommendation	Reference
1.	Dairy Effluent	Aerobic and Anaerobic technologies, Aerobic treatment particularly for high BOD.	50%	Membrane technology, Smart Monitoring and Control	[23]
2.	Beverage Effluent	Aeration Process, Tertiary Treatment, Aeration and pH Adjustment, sand Filtration, Activated Carbon Bed Filtration.	50%	Future Recommendation of beverage effluent Treatment. sequencing batch reactors.	[24]
3.	Sugar Effluent	Aerobic Treatment, Adsorption method, Anaerobic Treatment, Coagulation and flocculation used for effluent treatment, Electrocoagulation and flotation method, Thermolysis Treatment, Ceramic membrane treatment, oxidation processes.	74%	The Moving Bed Sequencing Batch Biofilm Reactor (MBSBR) treatment overcomes common issues associated with traditional treatment methods. It offers strong resistance to impact and eliminates the need for sludge return, making the system easier to operate.	[25]
4.	Chemical Effluent	Chemical Oxidation Processes such as ozonation /H_2O_2, UV photolysis/ H_2O_2, Photo-fenton processes, Ozonation Ultrasound system.	100%	Green and sustainable treatment methods such as phytoremediation, wetland systems, or constructed wetlands for the treatment of chemical effluents.	[26]
5.	Confectionery Effluent	anaerobic technology, aerobic treatment	88%	biological nutrient removal (BNR), enhanced biological phosphorus removal (EBPR)	[27]
6.	Food Industry effluents	Electrocoagulation Treatment anaerobic or aerobic biological treatment, coagulation-flocculation process, electrochemical processes, microbial fuel cells (MFCs), microalgae, water ultrafiltration, nanofiltration, and membrane technologies.	80%	ozone treatment, UV/H2O2.	[28]
7.	Sago effluent	Germination Treatment, complex filter comprising activated carbon and zeolite, aerobic treatment systems	90%	ultrafiltration (UF), nanofiltration (NF)	[29, 30]

S.No	Industry Effluents	Treatment Method Followed	Removal Rate %	Future Recommendation	Reference
8.	Sea Food effluent	Chemical precipitation, chemical oxidation ozonation, UV irradiation, halogenation, electrolysis, polymeric adsorption, granular activated carbon adsorption, dissolved air flotation (DAF), sludge treatments, Ozonation or UV irradiation.	90%	Chemical Coagulation and Flocculation	[31]
9.	Textile effluent	Photocatalytic oxidation, ozone, and Fenton's process. Adsorption, ion exchange, coagulation, and filtration.	98%	Carbon, biochar, or other adsorbents to remove residual dyes and organic compounds from textile effluents.	[32]
10.	Paper industry effluents	Ozonation, photo-Fenton	80%	photochemical reactions or catalytic ozonation	[33]

Table 2.
Review of Industrial Effluents Treatment technologies and recommendations.

practices in the dismantling of batteries, particularly in unorganized sectors; training and protective measures should be enforced to protect workers from exposure to hazardous materials [20]. Integrated recycling technology offers superior performance by achieving zero-pollution recycling of spent batteries. Biorecycling technology holds great potential for future development due to its advantages in energy savings, environmental protection, and high recycling efficiency through methods such as microbial and enzymatic degradation. Therefore, addressing the current challenges in waste battery recycling requires the development of new technologies and the enhancement of recycling systems, which are essential for the sustainable management of waste batteries [15]. Developing and adopting environmentally friendly recycling processes is paramount; transitioning from conventional hydrometallurgy to mildly acidic options and closed-loop processes will reduce health and environmental risks while maximizing resource recovery [22]. Automation in battery waste management can minimize human intervention, reducing health concerns and associated liabilities; although it may increase initial costs, the long-term benefits include improved efficiency and safety. Addressing these needs through strategic initiatives and technological advancements will significantly improve battery waste management, contributing to a healthier environment and sustainable resource utilization.

A comprehensive analysis of industrial effluent removal rates reveals considerable variation in treatment efficiencies across different sectors, highlighting the need for improvements in some industries to enhance environmental stability. It is shown in **Figure 1** that high-performing sectors such as chemical (100%), textile (98%), sago, and seafood (both 90%) benefit from advanced technologies like chemical oxidation, biological treatment, and membrane filtration, demonstrating effective environmental practices. In contrast, dairy and beverage industries lag behind, with lower removal rates of 50%, underscoring the need for enhanced treatment strategies. Moderate

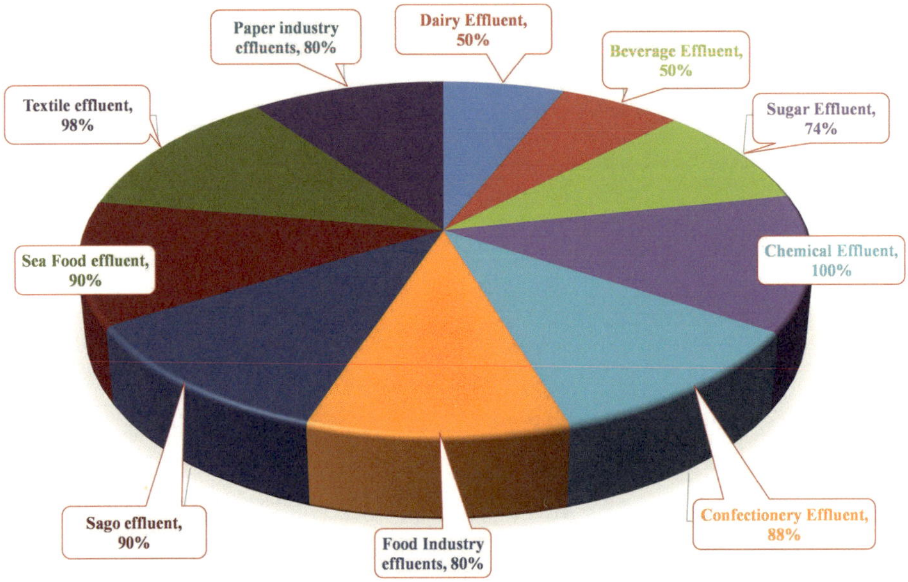

Figure 1.
The interpretation of the results reveals distinct trends in effluent removal efficiency across industries.

performers, including sugar (74%), confectionery (88%), food industry (80%), and paper industry (80%), show room for optimization through methods like anaerobic digestion and bio-filtration. This analysis underscores the importance of adopting more advanced and sustainable technologies, particularly for industries with lower efficiencies, such as aerobic/anaerobic digestion, microbial fuel cells (MFCs), and membrane filtration. Additionally, ongoing research and development (R&D), process optimization, and regular monitoring are essential to improving treatment processes and ensuring compliance with environmental standards, ultimately contributing to industrial sustainability.

4. Global industrial effluent management overview

Significant global efforts are focused on the remediation of textile wastewater for reuse. Researchers have explored various approaches, including the use of natural materials and microorganisms. For instance, coconut fiber has been utilized as a biomaterial in biofilm reactors to treat wastewater containing methyl orange, achieving successful decolorization using *Pseudomonas putida* under specific conditions [33]. Another study highlighted the potential of chickpea husk, an agricultural byproduct, to dye textiles such as cotton, wool, and silk without chemical auxiliaries, achieving high color depth and exhibiting antimicrobial properties against *E. coli* and *Staphylococcus aureus* [34].

Innovative technologies have been applied to treat synthetic textile wastewater as well. For example, a submerged membrane bioreactor (SMBR) has been shown to achieve high removal rates for biochemical oxygen demand (BOD), chemical oxygen demand (COD), and color [35]. Similarly, Upflow Anaerobic Sludge Blanket (UASB) reactors have been employed to recover methane as biogas, significantly reducing COD and chlorinated organic compounds from kraft industry effluents [36]. There is a growing demand for battery production due to increased energy requirements globally. Leading countries such as the US, UK, China, and India are expected to drive battery manufacturing by 2040. However, battery waste, particularly from alkaline and rechargeable batteries, poses serious environmental risks, as seen in the US and China [37]. Certain heavy metals are classified as xenobiotics due to their lack of beneficial roles in bodily functions and their harmful effects, even at low concentrations. Toxic metals such as cadmium, beryllium, aluminum, uranium, mercury, lead, bismuth, barium, antimony, and arsenic pose significant health risks. Elevated levels of these metal ions are particularly toxic to animals, including humans, as well as plants, and their water solubility is regarded as a major environmental concern [38]. Leather production, a significant economic contributor in countries like India, Brazil, and China, faces increasing environmental regulations. For example, the European Union has restricted hexavalent chromium, encouraging a shift toward sustainable practices.

Recent advancements in wastewater treatment involve innovative technologies such as membrane filtration, which effectively removes heavy metals from industrial effluents [39]. Adsorption, particularly using activated carbon, is a well-established method for eliminating organic pollutants and heavy metals [40]. Photocatalysis, which harnesses sunlight, is another eco-friendly solution, breaking down various organic and inorganic pollutants. Electrocoagulation and bioaugmentation are also gaining attention for their ability to treat complex industrial effluents by removing pollutants and enhancing biodegradation [41]. Nanotechnology offers promising solutions, especially through nanofiltration and the use of nanoparticles for adsorbing and removing contaminants. Ozonation, a powerful oxidation process, is highly effective in treating wastewater without leaving harmful byproducts [42]. For the treatment of wastewater with high ammonia concentrations, both biological and physicochemical methods are required. Stripping techniques are commonly used to remove ammonia gas, with the efficiency of the process being influenced by factors such as pH and temperature [43].

Conventional methods, like gravity separation and filtration, remain useful for treating oil-contaminated wastewater, although combining techniques often enhances efficiency [44]. Water scarcity and stringent regulations have made water reuse increasingly important in dye-intensive industries such as textiles and leather. Soluble dyes pose the greatest challenge among various dye types. Nanofiltration plays a crucial role in treating dye wastewater, but its primary limitation is membrane fouling. Coagulation and flocculation can significantly improve nanofiltration performance, aiding in water reuse and reducing fouling issues [45].

5. Future needs and recommendations based on current research

Current practices in industrial wastewater treatment often begin with screening and preliminary treatments to address clogging and smell issues by removing large solids and employing chemicals for neutralization. However, there is a pressing need to enhance these screening methods and improve smell control techniques to minimize secondary pollutants during the initial stages. Conventional centralized effluent treatment plants (CETPs) primarily rely on activated sludge systems, which can be ineffective in dealing with complex, slowly biodegradable compounds. Future research should focus on adopting more cost-effective and efficient alternative treatment technologies, such as Electrocoagulation and Fenton's Reagent, to address the challenges posed by recalcitrant organic compounds [46, 47].

Hybrid treatment approaches that combine technologies like phytosorption with adsorption or filtration with photocatalytic adsorption have shown potential. Moving forward, further development and integration of these methods should be explored to enhance efficiency, reduce maintenance costs, and ensure sustainability. In the realm of sophisticated simulation models, including fluid dynamics and artificial intelligence, ongoing research should aim to optimize these models to better predict and improve treatment processes. Developing more refined empirical and statistical models will be crucial in addressing challenges related to water demand and treatment efficiency [48].

Biotechnological solutions, such as bacterial cellulose membranes and natural polymeric flocculants, are being explored, yet there is a need to expand these technologies and advanced oxidation processes (AOPs) to handle complex wastewater mixtures. Future research should ensure that these methods are sustainable, biocompatible, and economically viable. For treating oily wastewater, integrating traditional and modern techniques is necessary to handle diverse particle compositions and complex mixtures. Research should focus on reducing energy costs and minimizing environmental impacts while enhancing treatment effectiveness.

When dealing with tannery effluents, techniques like electrochemical treatment and nanofiltration are used to manage high concentrations of toxic substances. Future efforts should aim to develop comprehensive and sustainable treatment solutions, focusing on reducing waste generation at the source and optimizing water reclamation processes [49–51]. Lastly, battery recycling technologies are being investigated with a focus on recovering metals from Li-ion batteries. Bridging the gap between bench-scale research and commercial application is essential, and collaborative efforts between academia and industry are needed to develop scalable and cost-effective recycling technologies [52, 53].

6. Regulatory frameworks for advancing industrial wastewater treatment technologies (environmental, public health, and resource management challenges)

- *Tighter discharge regulations:* Traditional CETPs often struggle to manage complex and slowly degradable pollutants, necessitating stricter discharge limits for persistent contaminants like heavy metals, organic pollutants, and emerging substances such as microplastics. This would encourage industries to implement advanced treatment technologies like Electrocoagulation, Fenton's Reagent, and hybrid systems.

- *Support for sustainable technologies:* Advanced oxidation processes (AOPs), biotechnological solutions, and integrated methods have shown potential for improving wastewater treatment. Regulatory frameworks should promote the use of these environmentally friendly technologies by providing incentives like subsidies, tax reliefs, or technical assistance for industries adopting sustainable solutions.

- Standardization of innovative approaches: As newer biotechnological solutions, such as bacterial cellulose membranes, and integrated treatment methods gain popularity, there is a need for regulatory standards to validate their effectiveness and safety at a large scale. Clearly defined guidelines will help in the proper implementation and oversight of these advanced techniques.

- *Mandates for water reuse:* With growing concerns over water scarcity, regulations should encourage or require the reuse of treated wastewater in industrial applications. This aligns with the review's recommendation to optimize water reclamation technologies, especially in industries like leather tanning and oil processing.

- *Waste minimization and resource recovery:* The review emphasizes the importance of minimizing waste and reclaiming resources. Regulatory frameworks should promote the adoption of circular economy practices, including waste reduction at the source and recovery of valuable byproducts through methods like nanofiltration and recycling.

- *Improved monitoring and enforcement:* For emerging contaminants, particularly those from battery waste, regulatory frameworks need to enforce rigorous monitoring and compliance measures. Periodic evaluations will help ensure industries adhere to updated standards, particularly as new technologies move from research to commercial-scale applications.

Overall, the review underscores that evolving regulatory frameworks are critical for advancing sustainable, cost-effective wastewater treatment technologies. These regulations must keep pace with technological innovations to address future environmental and resource management challenges effectively.

7. Conclusion and research recommendations

This review provides a thorough examination of both the progress and persistent issues in industrial wastewater treatment, emphasizing the vital role of initial processes like screening and smell control. Despite their significance, there is a pressing need for advancements in screening technologies and more effective smell control methods to reduce secondary pollution and improve preliminary treatment efficiency. The review highlights a notable shortfall in conventional Common Effluent Treatment Plants (CETPs), which often fail to address complex, slowly biodegradable pollutants. Advanced technologies such as Electrocoagulation and Fenton's Reagent offer promising alternatives for managing persistent organic pollutants, and future research should focus on developing these cost-effective solutions to overcome traditional system limitations. Hybrid treatment methods that integrate various technologies have demonstrated potential for enhancing treatment efficiency and sustainability, warranting further exploration and refinement of sophisticated simulation models to better address water demand challenges. Biotechnological solutions and advanced oxidation processes (AOPs) present opportunities for more sustainable and economically viable wastewater treatment, with particular emphasis on bacterial cellulose membranes and natural polymeric flocculants. The review also stresses the importance of combining traditional methods with modern techniques for treating oily wastewater and developing comprehensive solutions for tannery effluents to reduce waste and optimize water reclamation. Additionally, battery recycling technologies need further development to bridge the gap between research and commercial application, focusing on scalability and cost-effectiveness. In conclusion, while there have been significant advancements in wastewater treatment technologies, addressing existing research gaps and improving the effectiveness, sustainability, and economic viability of these methods remains crucial. Continued innovation and collaboration among academia, industry, and policymakers are essential for advancing wastewater treatment practices and tackling future environmental and resource management challenges.

Acknowledgements

The author acknowledges the use of ChatGPT for language polishing of the manuscript.

Author details

Murugesan Manikkampatti Palanisamy[1*], Padmapriya Veerappan[2], Kalaivani Ramanathan[2], Sankaranarayanan Devikarani[1], Dharunya Rajagopal[1], Kokila Kannusamy[1], Saranya Indrabal[1] and Anusha Kolanji[1]

1 Department of Food Technology, Excel Engineering College, Tamil Nadu, India

2 Department of Electronics Communication Engineering, Erode Sengunthar Engineering College, Tamil Nadu, India

*Address all correspondence to: engineermurugesh@gmail.com

References

[1] Prince OU, Ugochukwu E, Chibuzo VO. Environmental pollution: Causes, effects, and the remedies. Microorganisms for Sustainable Environment and Health. 2020;**1**(1):419-429. DOI: 10.1016/B978-0-12-819001-2.00021-8

[2] Srikanth. Wastewater effluent analysis for electric power plants. Water Science and Technology. 2015;**62**(10):2256-2262. DOI: 10.1186/s40643-017-0163-7

[3] Rahangdale RV, Kore SV, Kore VS. Waste management in lead acid battery industry: A case study. World Journal of Applied Environmental Chemistry. 2012;**1**:7-12

[4] Efremenkov VM. Liquid effluents released in nuclear industry. International Journal of Renewable Energy Technology Research. 2014;**3**(7):1-10. DOI: 10.1007/s10582-003-0075-y

[5] Kuyucak N. Mining, the environment and the treatment of mine effluents. International Journal of Environment and Pollution. 2008;**10**(2):315-325. DOI: 10.1504/IJEP.1998.005151

[6] Saha PD, Baskaran D, Malakar S, Rajamanickam R. Comparative study of biofiltration process for treatment of VOCs emission from petroleum refinery wastewater–A review. Environmental Technology and Innovation. 2015;**8**:441-461. DOI: 10.1016/j.eti.2017.09.007

[7] Noukeu NA, Priso RJ. Effluent treatment in food industry. Water Resources and Industry. 2016;**16**(1):1-18. DOI: 10.1016/j.wri.2016.07.001

[8] Raghunath BV, Punnagaiarasi A, Rajarajan G, Irshad A, Elango A. Impact of dairy effluent on environment—A review. In: Prashanthi M, Sundaram R, editors. Integrated Waste Management In India, Environmental Science and Engineering. Chem: Springer; 2016. pp. 239-249. DOI: 10.1007/978-3-319-27228-3_22

[9] Rajkumar K, Muthukumar M, Sivakumar R. Novel approach for the treatment and recycle of wastewater from soya edible oil refinery industry—An economic perspective. Resources, Conservation and Recycling. 2010;**54**(10):52-758. DOI: 10.26872/jmes.2018.9.1.3

[10] Mirunalini V, Mushahary J. Waste management in leather industry. International Journal of Civil Engineering and Technology. 2017;**8**(4):1394-1401

[11] Azizi E, Fazlzadeh M, Ghayebzadeh M, Hemati L, Beikmohammadi M, Ghafari HR, et al. Application of advanced oxidation process (H_2O_2/UV) for removal of organic materials from pharmaceutical industry effluent. Environment Protection Engineering. 2017:**43**(1):01-09. DOI: 10.5277/epe170115

[12] Levy GJ, Fine P, Bar-Tal A. Treated Wastewater in Agriculture. Blackwell Pub; 2011. p. 464. DOI: 10.1002/9781444328561

[13] Khandegar V, Saroha AK. Electrocoagulation for the treatment of textile industry effluent–A review. Journal of Environmental Management. 2013;**128**:949-963. DOI: 10.1016/j.jenvman.2013.06.043

[14] Hubbe MA, Metts D, Blanco ZK. Wastewater treatment and reclamation:

A review. Bio Resource. 2016;**11**(3):7953-8091. DOI: 10.2175/106143010X12851009156321

[15] Kang Z, Huang Z, Peng Q, Shi Z, Xiao H, Yin R, et al. Recycling technologies, policies, prospects, and challenges for spent batteries. iScience. 2023;**26**(11):108072

[16] Linzner R, Salhofer S. Municipal solid waste recycling and the significance of informal sector in urban China. Waste Management & Research. 2014;**32**(9):896-907. DOI: 10.1177/0734242X14543555

[17] Chagnes A, Pospiech B. A brief review on hydrometallurgical technologies for recycling spent lithium-ion batteries. Journal of Chemical Technology & Biotechnology. 2013;**88**(7):1191-1199

[18] Zhang S, Yu N, Zeng S, Zhou S, Zhang MS, Yu N, et al. An adaptive and stable bio-electrolyte for rechargeable Zn-ion batteries. Journal of Materials Chemistry A. 2018;**6**:12237-12243

[19] Tansel B. From electronic consumer products to e-wastes: Global outlook, waste quantities, recycling challenges. Environment International. 2017;**98**:35-45. DOI: 10.1016/j.envint.2016.10.002

[20] Vaverková MD. Landfill impacts on the environment—Review. Geosciences. 2019;**9**(10):431. DOI: 10.3390/geosciences9100431

[21] Brückner L, Frank J, Elwert T. Industrial recycling of lithium-ion batteries—A critical review of metallurgical process routes. Metals. 2020;**10**:1107. DOI: 10.3390/met10081107

[22] Porvali A, Wilson BP, Lundström M. Mechanical and hydrometallurgical processes in HCl media for the recycling of valuable metals from Li-ion battery waste. Waste Management. 2019;**142**:257-266. DOI: 10.1016/j.resconrec.2018.11.023

[23] Bhutiani R, Pratap H, Ahamad F, Kumar P, Kaushik PD. Efficiency assessment of effluent treatment plant (ETP) treating an automobile industry effluent (Sidcul) Haridwar. Environment Conservation Journal. 2017;**18**(1&2):95-102. DOI: 10.36953/ECJ.2017.181213

[24] Abdel-Fatah MA, Sherif HO, Hawash SI. Design parameters for waste effluent treatment unit for beverage production. Ain Shams Engineering Journal. 2017;**9**(5):1149-1155

[25] Sadaf S, Ayesha S, Fouzia Q, Shahid R, Abdul M. A perspective review on methodologies for treating sugar industry effluents. International Journal of Environmental Science and Natural Resources. 2019;**21**(4):117-122

[26] Ahmed MB, Zhou JL, Ngo HH, Guo W, Thomaidis NS, Xu J. Progress in the biological and chemical treatment technologies for emerging contaminant removal from wastewater: A critical review. Journal of Hazardous Materials. 2017;**323**(Pt. A):274-298. DOI: 10.1016/j.jhazmat.2016.04.0453

[27] Ozgun H, Karagul N, Dereli RK, Ersahin ME, Coskuner T, Ciftci DI, et al. Confectionery industry: A case study on treatability-based effluent characterization and treatment system performance. Water Science & Technology. 2012;**66**(1):15-20. DOI: 10.2166/wst.2012.094

[28] Qasim W, Mane AV. Characterization and treatment of selected food industrial effluents by coagulation and adsorption techniques. Journal of Water Resources and Industry. 2013;**1**(1):1-12

[29] Sivananda N. Impact of treated sago industry effluent on paddy. An

International Quarterly Scientific Journal. 2010;**9**:93-96

[30] Anbukumar S, Narad MP, Ashok Kumar M. Effluent treatment for sago industry using zeolite and activated carbon. Open Journal of Water Pollution and Treatment. 2014:**02**(01):18-26. DOI: 10.15764/WPT.2014.02003

[31] Venugopal V, Sasidhara A. Seafood industry effluents: Environmental hazards, treatment and resource recovery. Journal of Environmental Chemical Engineering. 2020;**9**(2):2213-3437

[32] Teshale A, Amare TA, Esayas A. Textile industry effluent treatment techniques, textile industry effluent treatment techniques. Journal of Chemistry. 2021;**2021**:1-14. Article ID 5314404

[33] Sen SK, Patra P, Das CR, Raut S, Raut S. Pilot-scale evaluation of biodecolorization and biodegradation of reactive textile wastewater: An impact on its use in irrigation of wheat crop. Water Resources and Industry. 2019;**21**:100106. DOI: 10.1016/j.wri.2019.100106

[34] Jose S, Pandit P, Pandey R. Chickpea husk–A potential agro waste for coloration and functional finishing of textiles. Industrial Crops and Products. 2019;**142**:111833. DOI: 10.1016/j.indcrop.2019.111833

[35] Niren P, Jigisha P. Textile wastewater treatment using a UF hollow-fibre submerged membrane bioreactor (SMBR). Environmental Technology. 2011;**32**(11):1247-1257. DOI: 10.1080/09593330.2010.534821

[36] Buzzini AP, Pires EC. Evaluation of aupflow anaerobic sludge blanket reactor with partial recirculation of effluent used to treat wastewaters from pulp and paper plants. Bioresource Technology. 2007;**98**:1838-1848. DOI: 10.1016/j.biortech.2006.06.030

[37] Zheng Z, Zhu Z, Lin X, Zhang Y, He Y, Cao H, et al. A mini-review on metal recycling from spent lithium-ion batteries. Engineering. 2018;**4**:361-370. DOI: 10.1016/j.eng.2018.05.018

[38] Amir W, Jahanzaib A, Farhat I, Ashif S, Zahid M, Ghulam M. Pollution status of Pakistan: A retrospective review on heavy metal contamination of water, soil, and vegetables. BioMed Research International. 2014;**2014**:1-29. Article ID 813206. DOI: 10.1155/2014/813206

[39] Gunatilake SK. Methods of removing heavy metals from industrial wastewater. Methods. 2015;**1**(1):14. DOI: 10.1007/s13201-020-01286-0

[40] Lakherwal D. Adsorption of heavy metals: A review. International Journal of Environmental Research and Development. 2014;**4**(1):41-48. DOI: 10.1155/2018/256803

[41] Syam BD, Anantha Singh TS, Nidheesh PV, Suresh Kumar M. Industrial wastewater treatment by electrocoagulation process. Separation Science and Technology. 2020;**55**(17): 3195-3227. DOI: 10.3390/w12123426

[42] Ledakowicz S, Żyłła R, Paździor K, Wrębiak J, Sójka-Ledakowicz J. Integration of ozonation and biological treatment of industrial wastewater from dyehouse. Ozone Science and Engineering. 2017;**39**(5):357-365. DOI: 10.1016/j.cej.2018.10.093

[43] Farghali M, Chen Z, Osman AI, Israa MA, Dalia H, Ikko I, et al. Strategies for ammonia recovery from wastewater: A review. Environmental Chemistry Letters. 2024;**1**(1):01-53. DOI: 10.1007/s10311-024-01768-6

[44] Galdino CJS, Maia ADM, Meira HM, Souza TC, Amorim JDP, Almeida FCG, et al. Use of a bacterial cellulose filter

for the removal of oil from wastewater. Process Biochemistry. 2020;**91**:288-296

[45] Zahrim AY, Tizaoui C, Hilal N. Coagulation with polymers for nanofiltration pre-treatment of highly concentrated dyes. Desalination. 2011;**266**(1-3):1-16. DOI: 10.1016/j.desal.2010.08.012

[46] Lopez A, Ricco G, Mascolo G, Tiravanti G, Di Pinto AC, Passino R. Biodegradability enhancement of refractory pollutants by ozonation: A laboratory investigation on an azo-dyes intermediate. Water Science and Technology. 1998;**384**(4-5):239-245

[47] Kumar MD, Tortajada C. Assessing Wastewater Management in India. Singapore: Springer; 2020. pp. 23-30

[48] Medeiros ADMd, Silva Junior CJGd, Amorim JDPd, Durval IJB, Costa AFdS, Sarubbo LA. Oily wastewater treatment: Methods, challenges, and trends. Processes. 2022;**10**(4):743. DOI: 10.3390/pr10040743

[49] Anjum MN, Rasheed HU, Ahmed W. Impact of waste water treatment on quality of infuent and effluent water. Imperial Journal of Interdisciplinary Research (IJIR). 2016;**1**(1):844-849

[50] Sathya K, Nagarajan K, Carlin Geor Malar G, Rajalakshmi S, Raja Lakshmi G. Comprehensive review on comparison among effluent treatment methods and modern methods of treatment of industrial wastewater effluent from different sources. Applied Water Science. 2022a;**12**(4):58-70. DOI: 10.1007/s13201-022-01594-7

[51] Orozco, Guillén AO, Velazquez Silva RI, Moguel González B, Guell YA, Garciadiego Fossas P, Custodio Gómez IG, et al. Acute IgA-dominant glomerulonephritis associated with syphilis infection in a pregnant teenager: A new disease association. Journal of Clinical Medicine. 2019;**8**(1):114 (01), 1-10. DOI: 10.3390/jcm8010114

[52] Mänttäri M, Kuosa M, Kallas J, Nyström M. Membrane filtration and ozone treatment of biologically treated effluents from the pulp and paper industry. Journal of Membrane Science. 2008;**309**:112-119. DOI: 10.1016/j.memsci.2007.10.019

[53] Sathya K, Nagarajan K, Carlin Geor Malar G. A comprehensive review on comparison among effluent treatment methods and modern methods of treatment of industrial wastewater effluent from different sources. Applied Water Science. 2022b;**12**(1):70-78. DOI: 10.1007/s13201-022-01594-7

Section 2

Waste Recovery, Recycling, and Bio-Waste Management

Chapter 3

Citrus Peels an Effective Sources of Bioactive Compounds

Oyebola Odunayo Olabinjo

Abstract

Citrus fruits, a cornerstone of global agriculture, generate substantial byproducts in the form of citrus peels and others. Traditionally considered as waste, recent studies have unveiled a rich reservoir of bioactive compounds within citrus peels. Citrus peels accounting to over 40% of the total mass of citrus fruits. The most common citrus fruits in the World is sweet orange (*Citrus sinesis*) with peels ranging between 45 and 50% of the total mass of the fruits, rich in protein of 7.5% and crude fiber of 12.7% which can be used as an additive in processed foods. The citrus peels are sources of extracting essential oil of about 49.3% and natural antioxidant with different chemical (limonene, myrene and octanol acetate) compounds. The extensive study of the chemical and bioactive constituents of citrus peels has yielded valuable insights with significant implications for various industries including cosmetics, food industry and pharmaceutical industries. Through rigorous analysis, we have elucidated the presence of key compounds including flavonoids, limonoids, carotenoids, and essential oils. The use of citrus peels will promote sustainable disposal of orange peels and reduce pollution. The bioactive substances reduced ROS, lowering the risk of metabolic syndrome, hyperglycemia, cancer, heart disease, and neurological illnesses.

Keywords: bio-active compounds, nutritional, citrus peels, antioxidant, essential oil

1. Introduction

Citrus fruits, integral to global agriculture, are renowned for their culinary, medicinal, and esthetic attributes [1]. Citrus (*Citrus spp*) stands as one of the most abundant fruit crops globally, contributing to an estimated yearly production of 124.24 million tons [2]. Within the extensive citrus family, prominent members include sweet oranges (*Citrus sinensis L.*), lemons (*Citrus limon L.*), limes (*Citrus aurantifolia L.*), tangerines/mandarins (*Citrus reticulata L.*), grapefruit (*Citrus paradisi L.*), pomelos (*Citrus maxima L.*) and other hybrids. Citrus peels are rich in a range of functional compounds, according to phytochemical and nutritional analyses. These include flavonoids [3–5], carotenoids [6–8], limonoids, essential oil [9], pectin [10, 11]. Several other bioactive components with excellent antioxidant [3], and health-promoting potential [12–15] are also present in citrus peels. These compounds may find application in the cosmetics, food processing, agricultural production and pharmaceutical industries [16]. Studies revealed that about 140 different chemical components have been isolated and identified from citrus peels. The distinct fragrance of citrus fruits is attributed in part to flavonoids and limonoids present in their peels [17]. Furthermore, citrus fruits serve as significant sources of bioactive compounds, including antioxidants like ascorbic acid, flavonoids, and phenolic compounds, essential for human nutrition [18] and health. Citrus fruits also provide folic acid, vitamin B (thiamine), potassium, phosphorus, calcium, iron, magnesium, sodium, and sulfur [19]. This chapter delves into a thorough exploration of the chemical composition of citrus peels and explores potential applications for the bioactive compounds in diverse industries such as pharmaceuticals, food processing, cosmetics, and agriculture.

Waste is defined as anything no longer useful to the disposer or any material resulting from an activity with no immediate economic demand that must be disposed [20]. In Nigeria, as of 2007, animal and agricultural wastes amounted to 0.781 and 0.256 million tons per day, respectively [21]. However, waste from agricultural processing presents a promising source of bioactive compounds and non-bulky caloric fiber suitable for technological or nutritional purposes [21]. Citrus peels, the primary waste from concentrated citrus juice processing, are globally abundant, representing the most common fruit worldwide. Despite being considered waste, citrus peels can serve as raw materials for producing volatile oils, dietry fiber and compounds with antioxidant properties. The Raw Materials Research and Development Council (RMRDC) notes a local demand for volatile oil exceeding 100,000 kg annually, a demand that local production efforts could meet.

1.1 Composition of citrus fruits

Citrus fruits exhibit a round, textured, and green to yellow skin, measuring approximately 20–30 cm in length. They are encased in a robust peel, also known as epicarp or flavedo, which serves as a protective cover shielding the fruit from environmental influences. The endocarp of these fruits is rich in soluble sugars and contains significant quantities of vitamin C, pectin, fibers, various organic acids, and potassium salt, collectively contributing to the distinctive "citrus flavor" [22]. **Figure 1** illustrates the cross-section of a citrus fruit and its peels. In Africa, the total production of various citrus fruit varieties is 3,741,000 tonnes, with Nigeria contributing significantly at 3,240,000 tonnes [23].

During the citrus fruit processing, a substantial byproduct takes the form of citrus peels, traditionally viewed as waste material [24]. The peels, derived from the pericarp of fresh fruits, are generated in large quantities during the citrus season, posing environmental challenges as they are not utilized productively. Despite the significant contribution of the citrus industry to the global economy through the production of a

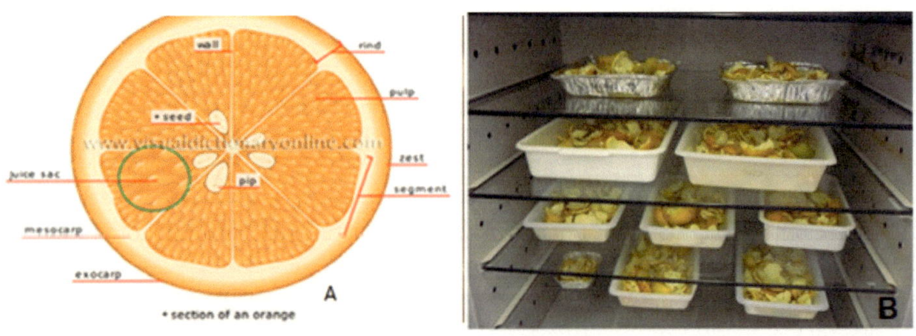

Figure 1.
Section of citrus and peels of citrus.

large quantity of fruit each year [25], recent research has uncovered a valuable reservoir of bioactive compounds within these seemingly discarded peels. The exploration of citrus peels stems from their potential as a robust source of bioactive compounds, garnering attention for their diverse health benefits. These compounds encompass flavonoids, limonoids, carotenoids, and essential/volatile oils [26]. Research revealed that the peel of most citrus fruits has higher levels of antioxidants and polyphenols than the edible pulp [7, 27]. Therefore, it is necessary to extract these health- beneficial components from citrus fruit peels. Furthermore, citrus peel is a viable substitute for lignocellulosic biomass in the production of biofuels due to its low lignin content [28].

1.2 Nutrient composition of sweet orange (*Citrus sinensis*) peels

The nutritional composition of sweet orange peels, as documented by Olabinjo et al. [29], reveals a crude protein content of 7.15% and crude fiber content of 12.79%. The high crude protein content suggests their potential as substantial protein supplements, contributing to the synthesis of hormones that regulate various bodily functions like growth, repair, and protein maintenance. With a crude fiber content of 12.79%, it offers a rich source of dietary fiber, surpassing the value found in water leaf (*Talinum triangulare*) at 12%, as reported by Aja et al. [30]. Dietary fiber aids in avoiding hydrolysis, digestion, and absorption in the human small intestine, helping to maintain insulin levels and reduce pre-prandial cholesterol levels [31]. The recommended daily intake of dietary fiber is 21–25 g for women and 30–38 g for men [31], a requirement that can be met with citrus peels as a supplement in the human diet, given their obtained percentage of crude fiber. Health-conscious individuals often prefer natural supplements, fearing potential toxicity from synthetic ingredients. Fiber-rich byproducts and bioactive compounds are considered valuable assets for food processors. Incorporating dietary fiber from citrus peels into the diet can result in safer and more economical food options with multiple health benefits. Nassar et al. [32], proposed the inclusion of 15% orange peel and pulp in biscuit production, emphasizing their suitability as a source of dietary fiber along with associated bioactive compounds like flavonoids and carotenoids. These peels also contain proteins, crude fiber, and various minerals. The nutritional composition analysis indicates sweet orange as a promising source of proteins and crude fiber for use in food processing. Dietary fiber consumption is important for the prevention, reduction, and treatment of chronic diseases like cancer, obesity, diabetes, cardiovascular disease, and gastrointestinal disorders [33]. Dietary fiber supplements can produce cheaper, safer foods with a host of health advantages, promoting physiological functions like lowering blood triglycerides and glucose control.

1.3 Citrus peels essential oil

The citrus peel comprises the outer layer called the exocarp, which includes irregular thin-walled cells known as pearenchymatous. This layer encloses numerous glands or oil sacs, representing the source of citrus essential oil (EO) obtained as a by-product of citrus processing [34–36]. Citrus fruits are recognized not only for their culinary use but also as excellent reservoirs of essential oils. Citrus EO has become significant in the food industry due to its antimicrobial properties against bacteria and fungi [35, 37, 38]. Essential oils, being volatile compounds, contribute to the distinctive aroma and flavor of citrus fruits. Limonene, a major component of citrus essential oils, has garnered extensive attention for its diverse applications. With antimicrobial properties, limonene is considered a potential natural preservative in the food industry and has been explored for therapeutic purposes, including its anti-anxiety potential and use in aromatherapy.

Essential oil is predominantly found in the peels of citrus species compared to other parts and finds wide applications in the food industry as an additive, nutritional supplement, and in various industrial processes [39]. Citrus species offer a variable oil content suitable for edible and industrial applications. There is a growing global interest in the safe extraction of natural bio actives from industrial waste and by-products, such as plant peels, for functional foods and nutraceutical applications [40]. Traditional methods like distillation (hydro, steam, and destructive), maceration, and expression are commonly used for essential oil extraction from plant material [41]. However, to address limitations such as extraction time, cost, and potentially improve yield and quality, newer techniques like microwave-assisted extraction (MAE), pressurized liquid extraction (PLE), supercritical fluid extraction, and ultrasound-assisted extraction have been developed [42].

1.4 Essential oil yield from citrus species

The essential oil yields documented in various studies for *citrus sinensis* peels exhibit a range of values. Franco-Vega et al. [43] reported yields ranging from 0.92% to 2.73%, Kamal et al. [44] produced oils within the range of 0.24–1.07%, while Bourgou et al. [45] obtained a yield of 0.74%. Olabinjo et al. [29] reported essential oil yields between 21.9% and 49.3%, and Ademosun et al. [46] achieved a yield of 1.54%. Lime peels (*Citrus latifolia* Tanaka) yielded 5.45% essential oil using a method similar to Attisantos et al. [47]. Mercy et al. [48] introduced an improved distillation method, resulting in essential oil yields of 4.23% for *Citrus sinensis* peels and 5.865% for *Citrus reticulata* peels. Franco-Vega et al. [43] also reported orange peel extract yields ranging from 0.92% to 2.73%. Megha and Mumtaj [49] reported a 1.16% yield for sweet lime using microwave-assisted hydrodistillation. Ahmad et al. [50] (2006) documented essential oil yields varying from 0.30% to 1.21% for four citrus varieties from Pakistan. Kamal et al. [44] highlighted that *C. sinensis* had the highest oil value yield of 0.24% to 1.07%, followed by *C. reticulata* with 0.30% to 0.50%, and the least was *C. paradisii* with 0.20% to 0.40%. This study shows significant variations in essential oil yields compared to the literature, suggesting potential influences from factors such as climatic and environmental conditions, soil variations, seasonal fluctuations, geographical location, the stage of the vegetative cycle, and the extraction method employed [51].

1.5 Chemical and bioactive composition of citrus peels

Plant based medicines are used worldwide and have always played a key role in health care systems. It contains biologically active chemical substances such as vitamin C, flavonoids, limonoids and etc. Understanding the bioactive potential of citrus peels not only provides opportunities for innovative product development, but also connects with global efforts in the agricultural and food processing sectors toward sustainability and waste reduction [52].

We may contribute to a circular economy model by repurposing citrus peels, reducing environmental impacts connected with waste disposal. Citrus peels are well-known for their rich and diverse chemical composition, which includes a wide range of bioactive compounds. The primary constituents of these include flavonoids, limonoids, carotenoids, and essential oils.

a. *Flavonoids*: citrus peels contain a high concentration of polyphenolic compounds. These compounds contribute not only to the vibrant pigmentation of citrus fruits but also to their potential health benefits. Citrus flavonoids, such as hesperidin and naringin, have gained attention for their potent antioxidant properties. Citrus flavonoids like hesperidin and naringin have gained popularity due to their powerful antioxidant effects. These antioxidants are important in scavenging free radicals, lowering oxidative stress and protecting cells from damage. Flavonoids have also been shown to have anti-inflammatory properties, making them potential candidates for alleviating chronic inflammatory diseases. Flavonoids, for example, have antioxidant and anti-inflammatory properties that have been linked to a lower risk of chronic diseases such as cardiovascular disease and some types of cancer [53].

b. *Limonoids*: this is another group of bioactive compound found in citrus peels have shown promise in cancer prevention and treatment. These triterpenoid compounds possess anti-cancer properties and have exhibited inhibitory effects on various stages of cancer development. Limonoids, such as limonin and nomilin, have been studied for their potential to inhibit tumor growth and suppress the proliferation of cancer cells. Limonoids, another group of bioactive compounds, have demonstrated promising anti-cancer properties [54]. Thousands of tonnes of by-products are created during the citrus juice extraction process [55], and the food industry has shown special interest in finding the utilization for citrus industry by-products.

The examination of retention characteristics and GCMS analysis disclosed the identification of 47 constituents distributed among six groups, with terpenes comprising the majority (95.13%), followed by aldehydes (1.19%), alcohols (0.68%), esters (1.65%), oxide (0.1%), and ketone (0.05%) from Noot Ketone. These constituents collectively accounted for approximately 99.8% of the essential oil. The research findings highlighted the predominant constituents in citrus sinesis peels, with limonene (90.72%) taking the lead, accompanied by myrcene (2.82%), octanol acetate (1.24%), nonanal (0.58%), sabinene (0.39%), and elemol (0.14%). Previous studies by Khaoula et al. [56], and Olabinjo et al. [57] have indicated that essential oil constituents generally fall into categories such as terpenes (monoterpenes and sesquiterpenes), aromatic compounds (aldehyde, alcohol, phenol, methoxy derivatives), and terpenoids (isoprenoids). The research results also suggest the potential utility of *Citrus sinesis* as a limonene source for applications in nutrition, pharmaceuticals, and cosmetics, as reported by Vivian et al. [58]. Additionally, other compounds, including myrcene (2.82%), octanol acetate (1.24%), nonanal (0.58%), sabinene (0.39%), and elemol (0.14%), were detected at lower levels.

c. *Carotenoids*: this contribute to the vibrant colors of citrus fruits, ranging from yellow to orange and red hues. These compounds are well-known for their antioxidant capabilities, which play a crucial role in protecting cells from oxidative damage. The most prevalent carotenoid in citrus peels is β-carotene, a precursor of vitamin A. This compound not only contributes to the nutritional value of citrus fruits but also offers potential health benefits, particularly in maintaining healthy vision and supporting immune function.

1.6 Health benefit of citrus peels

Citrus peels, which are often disregarded, are rich in various bioactive compounds that have numerous health advantages. They contain essential nutrients, such as vitamins, minerals, and dietary fiber, as well as powerful antioxidants. Citrus bioactive compounds, particularly flavonoids, carotenoids, terpenes, and limonoids, have the antioxidant ability to reduce disorders associated with oxidative stress [59, 60]. As a result, they may be useful in the treatment of obesity [61], inflammatory diseases [62–64], atherosclerosis [65], neurodegenerative diseases [66–68], and cancer [69, 70]. The following are some of the health benefits:

a. *Rich source of dietary fiber*: citrus peels are a notable source of dietary fiber, which aids in digestion and promotes regular bowel movements [71]. The fiber content can help prevent constipation and promote a healthy gastrointestinal system.

b. *Rich in vitamins*: citrus peels, especially from fruits like oranges and lemons, are packed with vitamins such as vitamin C, vitamin A, and various B-complex vitamins [72]. Vitamin C, for instance, is crucial for the immune system and acts as a potent antioxidant.

c. *Antioxidant properties*: citrus peels are rich in flavonoids, limonoids, and carotenoids, which possess potent antioxidant properties [73]. These antioxidants help combat oxidative stress, reducing the risk of chronic diseases like heart disease, diabetes, and certain types of cancer. These variations in the antioxidant activity between extracts can be explained by the differential solubilization of antioxidant compounds, as was reported for citron blood orange [74]. The antioxidant capacity of citrus peels might be related to the presence of phenolic compounds and flavonoids. Multiple compounds such as flavanones, flavanone glycosides and polymethoxylated flavones are special to citrus, which are comparatively uncommon in other plants [75]. Mandarin essential oil exhibited the highest 2,2′-Azino-bis(3-ethylbenzothiazoline-6-sulfonic acid (ABTS) radical cation reducing activities and ferric reducing antioxidant power (FRAP) in a comparative study between essential oils extracted from grapefruit, lemon, orange, and mandarin. On the other hand, lemon essential oil demonstrated the highest 2, 2-diphenyl-1-picrylhydrazyl (DPPH) free radical scavenging and cupric ion reducing antioxidant capacity (CUPRAC) [76].

d. *Anti-inflammatory effects*: citrus peel compounds, such as hesperidin and naringin, have anti-inflammatory effects [77]. These compounds can help reduce inflammation in the body, potentially benefiting individuals with conditions like arthritis. It has been demonstrated that the bioactives in citrus fruit juices directly inhibit virus growth. Hesperidin inhibited influenza virus replication and advancement in human lung cancer A549 cells by upregulating the p38 signaling pathway, according to Dong et al. [78]. It has been demonstrated that hesperetin, naringenin, and hesperidin inhibit important proteases involved in coronavirus replication and stop the virus from entering host cells [61, 79, 80].

e. *Cholesterol management*: certain compounds in citrus peels, particularly polymethoxylated flavones (PMFs), have been shown to have cholesterol-lowering effects [81]. This can contribute to cardiovascular health by reducing the levels of LDL cholesterol in the blood.

f. *Blood sugar regulation*: studies have indicated that citrus peel extracts may help regulate blood sugar levels, making them potentially beneficial for individuals with diabetes [82]. The presence of flavonoids and other bioactive compounds is believed to play a role in this effect.

g. *Weight management*: the fiber content in citrus peels can contribute to a feeling of fullness, potentially aiding in weight management efforts [83]. Adding citrus peels in the diet may help control appetite and prevent overeating.

h. *Anti-microbial properties*: citrus peel extracts have demonstrated antimicrobial activity against various pathogens, including bacteria and fungi [84]. This suggests a potential role in supporting the immune system and preventing infections.

i. Citrus pectin and coumarin, in addition to flavonoids, carotenoids, terpenes, and limonoids, have positive health effects [65, 85]. For example, citrus pectin oligosaccharides and their microbial metabolites inhibited the production of proinflammatory mediators (TNF-α, IL-6, IL-10, and NF-κB mRNA) in LPS-treated human macrophages, hence exhibiting anti-atherosclerosis properties [65]. It is generally accepted (GRAS) that limonene and other terpene-rich citrus flavoring compounds, including oil, essential oil, whole fruit extract, and peel extract, are safe [86].

2. Conclusion

The extensive study of the chemical composition and bioactive compounds of citrus peels has yielded valuable insights with significant implications for various industries including cosmetics, food industry and pharmaceutical industries. Through rigorous analysis, we have elucidated the presence of key compounds including flavonoids, limonoids, carotenoids, and essential oils. These compounds contribute not only to the sensory characteristics of citrus fruits but also hold promise for numerous applications. The abundance of flavonoids in citrus peels, such as hesperidin and naringin, highlights their potential as natural antioxidants. These antioxidants play an important role in the prevention of chronic illnesses by mitigating oxidative stress. The versatility of citrus peel extracts extends beyond the boundaries of health and wellness. The essential oils derived from citrus peels, particularly limonene, have showcased antimicrobial properties, holding promise for applications in food preservation and healthcare.

Author details

Oyebola Odunayo Olabinjo
Department of Agricultural and Environmental Engineering, School of Engineering and Engineering Technology, Federal University of Technology, Akure, Nigeria

*Address all correspondence to: ooolabinjo@futa.edu.ng; binjo_joko@yahoo.co.uk

References

[1]] Liu Y, Heying E, Tanumihardjo SA. History, global distribution, and nutritional importance of citrus fruits. Comprehensive Reviews in Food Science and Food Safety. 2012;**11**(6):530-545. DOI: 10.1111/j.1541-4337.2012.00201.x

[2] FAO, IFAD, UNICEF, WFP and WHO. The State of Food Security and Nutrition in the World. Food and Agriculture Organization of the United Nations Rome; 2019. ISBN: 978-92-5-131570-5. Available from: www.fao.org/publications, www.fao.org/contact-us/licence-request

[3] Long XY, Zeng XG, Yan HT, Xu MJ, Zeng QT, Xu C, et al. Flavonoids composition and antioxidant potential assessment of extracts from Gannanzao Navel Orange (*Citrus sinensis* Osbeck Cv. Gannanzao) peel. Natural Product Research. 2021;**35**:702-706

[4] El-Kersh DM, Ezzat SM, Salama MM, Mahrous EA, Attia YM, Ahmed MS, et al. Anti-estrogenic and anti-aromatase activities of citrus peels major compounds in breast cancer. Scientific Reports. 2021;**11**:7121

[5] Chen Y, Pan HL, Hao SX, Pan DM, Wang GJ, Yu WQ. Evaluation of phenolic composition and antioxidant properties of different varieties of Chinese citrus. Food Chemistry. 2021;**364**:130413

[6] Saini A, Panesar PS, Bera MB. Valuation of *Citrus reticulata* (kinnow) peel for the extraction of lutein using ultrasonication technique. Biomass Conversation Biorefinery. 2021;**11**:2157-2165

[7] Costanzo G, Iesce MR, Naviglio D, Ciaravolo M, Vitale E, Arena C. Comparative studies on different citrus cultivars: A revaluation of waste mandarin components. Antioxidants. 2020;**9**:517

[8] Murador DC, Salafia F, Zoccali M, Martins PLG, Ferreira AG, Dugo P, et al. Green extraction approaches for carotenoids and esters: Characterization of native composition from orange peel. Antioxidants. 2019;**8**:613

[9] Singh B, Singh JP, Kaur A, Yadav MP. Insights into the chemical composition and bioactivities of citrus peel essential oils. Food Research International. 2021;**143**:110231

[10] Nuzzo D, Picone P, Giardina C, Scordino M, Mudò G, Pagliaro M, et al. New neuroprotective effect of lemon IntegroPectin on neuronal cellular model. Antioxidants. 2021;**10**:669

[11] Colodel C, Vriesmann LC, Teofilo RF, Petkowicz CLD. Extraction of pectin from ponkan (*Citrus reticulata* Blanco cv. Ponkan) peel: Optimization and structural characterization. International Journal of Biological Macromolecules. 2018;**117**:385-391

[12] Mahato N, Sharma K, Sinha M, Cho MH. Citrus waste derived nutra−/pharmaceuticals for health benefits: Current trends and future perspectives. Journal of Functional Foods. 2018;**40**:307-316

[13] Liu N, Li X, Zhao P, Zhang XQ, Qiao O, Huang LQ, et al. A review of chemical constituents and health-promoting effects of citrus peels. Food Chemistry. 2021;**365**:130585

[14] Gogoi M, Boruah JLH, Bora PK, Das DJ, Famhawite V, Biswas A, et al. *Citrus macroptera* induces apoptosis

via death receptor and mitochondrial mediated pathway as prooxidant in human non-small cell lung cancer cells. Food Bioscience. 2021;**43**:101293

[15] Abdelghffar EA, El-Nashar HAS, Al-Mohammadi AGA, Eldahshan OA. Orange fruit (*Citrus sinensis*) peel extract attenuates chemotherapy-induced toxicity in male rats. Food & Function. 2021;**12**:9443-9455

[16] Hou Y, Dan X, Babbar M, Wei Y, Hasselbalch SG, Croteau DL, et al. Ageing as a risk factor for neurodegenerative disease. Nature Reviews. Neurology. 2019;**15**(10):565-581

[17] Manthey JA. Fractionation of orange peel phenols in ultra-filtered molasses and balance studies of their antioxidant levels. Journal of Agricultural and Food Chemistry. 2004;**52**(25):7586-7592

[18] Kamran G, Youcef G, Ebrahimzadeh MA. Antioxidant activity, phenol and flavonoid contents of 13 citrus species peels and tissues. Pakistan Journal of Pharmaceutical Sciences. 2009;**22**(3):277-281

[19] Nagy V, Szabo Z, Mark L, Deli J. Comparative study on the carotenoid citrus. Innovative Food Science and Emerging Technologies. 2007;**8**:390-394

[20] Nigeria Institute of Safety Professionals – NISP. Contractor Employee HSE Training Manual Level 3. Harcourt: ECNEC Ltd. 2003

[21] Sambo AS. Renewable energy development in Nigeria: A situation report. A keynote address presented at the International Workshop on Renewable Energy f or sustainable Development in Africa, IWRESDA 30th July to 1st August. 2007

[22] Cava R, Ladero L. Pomegranate peel as a source of antioxidants for the control of lipid and protein oxidation during the ripening of Iberian dry uncured sausages. Meat Science. 2023;**202**:109198. DOI: 10.1016/j.meatsci.2023.109

[23] Ezejiofor TLN, Eke NV, Okechukwu RI, Nwoguikpe RN, Duru CM. Waste to wealth: Industrial raw materials potential of peels of Nigerian sweet orange (*Citrus sinensis*). African Journal of Biotechnology. 2011;**10**(33):6257-6264

[24] FAO. FAOSTAT data 2005. Food and Agricultural Organisation of the United Nations, 00100, Rome, Italy. February 2005. 2004

[25] Oboh G, Akinyemi AJ, Ademiluyi AO. Antioxidant and inhibitory effects of aqueous extracts of some sweet orange varieties on key enzymes relevant to type-2 diabetes in vitro. Journal of Food Composition and Analysis. 2006;**19**(4):372-376

[26] Ladaniya MS. Citrus Fruit: Biology, Technology and Evaluation. Atlanta, USA: Elsevier Inc., Academic Publisher; 2008. pp. 1-10. DOI: 10.1016/B978-012374130-1.50003-6

[27] Singh B, Singh JP, Kaur A, Singh N. Phenolic composition, antioxidant potential and health benefits of citrus peel. Food Research International. 2020;**132**:109114

[28] Jeong D, Park H, Jang BK, Ju YB, Shin MH, Oh EJ, et al. Recent advances in the biological valorization of citrus peel waste into fuels and chemicals. Bioresource Technology. 2021;**2021**(323):124603

[29] Olabinjo OO, Ogunlowo AS, Ajayi OO, Olalusi AP. Analysis of physical and chemical composition of sweet Orange (*Citrus sinensis*) peels. International Journal of Environment,

Agriculture and Biotechnology (IJEAB). 2017;**2**(4):2201-2206

[30] Aja PM, Okaka ANC, Ibiam UA, Uraku AJ, Onu PN. Proximate analysis of Talinumtriangulare (Water Leaf) leaves and its softening principle. Pakistan Journal of Nutrition. 2010;**9**(6):524-526

[31] Food and Nutrition Board, Institute of Medicine. Dietary Reference Intakes. Proposed Definition of Dietary Fibre. A Report of the Panel on the Definition of Dietary Fibre and the Standing Committee on the Scientific Evaluation of Dietary Reference Intakes. Washington, DC: National Academy Press; 2001

[32] Nassar AG, AbdEl-Hamied AA, El-Naggar EA. Effect of citrus by-products flour incorporation on chemical, rheological and organoleptic characteristics of biscuits. World Journal of Agricultural Science. 2008;**4**(5):612-616

[33] Figuerola F, Hurtado ML, Estevez AM, Chiffelle I, Asenjo F. Fibre concentrates from apple pomace and citrus peel as potential fibre sources for food enrichment. Food Chemistry. 2005;**91**:395-401

[34] Farhat A, Fabiano-Tixier AS, El Maataoui M, Maingonnat JP, Romdhane M, Chemat F. Microwave steam diffusion for extraction of essential oil from orange peel: Kinetic data, extract's global yield and mechanism. Food Chemistry. 2011;**125**(1):255-261

[35] Velazquez-Nunez M, Avila-Sosa R, Palou E, LopezMalo A. Antifungal activity of orange (*Citrus sinensis* var. Valencia) peel essential oil. Food Contamination. 2013;**31**:1-4

[36] Lago S, Rodríguez H, Arce A, Soto A. Improved concentration of citrus essential oil by solvent extraction with acetate ionic liquids. Fluid Phase Equilibria. 2014;**361**:37-44

[37] Bousbia N, Vian M, Ferhat M, Meklati B, Chemat F. A new process for extraction of essential oil from citrus peels: Microwave hydro diffusion and gravity. Journal of Food Engineering. 2009;**90**:409-413

[38] Rezzoug SA, Louka N. Thermomechanical process intensification for oil extraction from orange peels. Innovative Food Science and Emerging Technologies. 2009;**10**:530-536

[39] Maria CC, Rubria ER, Jose EB, Gloria MM, Jose LN, Hugo J. Characterization of volatile compounds in the essential oil of sweet lime. Chilean Journal of Agricultural Research. 2012;**72**(2):275-282

[40] Borah MS, Tiwari A, Sridhar K, Narsaiah K, Nayak PK, Stephen Inbaraj B. Recent trends in valorization of food industry waste and by-products: Encapsulation and in vitro release of bioactive compounds. Food. 2023;**12**:3823. DOI: 10.3390/foods12203823

[41] Stahl-Biskup E, Saez F. Thyme the Genus Thymus. New York: Taylor and Francis; 2002

[42] Wang L, Weller LC. Recent advances in extraction of nutraceuticals from plants. Trends in Food Science and Technology. 2006;**1**:300-312

[43] Franco-Vega A, Ramirez-Corona N, Palou E, LópezMalo A. Estimation of mass transfer coefficients of the extraction process of essential oil from orange peel using microwave assisted extraction. Journal of Food Engineering. 2016;**170**:135-141

[44] Kamal GM, Anwar F, Hussain AI, Sarri N, Ashraf MY. Yield and chemical composition of Citrus essential oils as affected by drying pre-treatment of peels. International Food Research Journal. 2011;**18**(4):1275-1282

[45] Bourgou SF, Rahali Z, Ourghemmi I, Tounsi MS. Changes of peel essential oil composition of four tunisian citrus during fruit maturation. The Scientific World Journal. 2012: 528593

[46] Ademosun AO, Oboh G, Olupona AJ, Oyeleye SI, Adewuni TM, Nwanna EE. Comparative study of chemical composition, in vitro inhibition of cholinergic and monoaminergic enzymes, and antioxidant potentials of essential oil from peels and seeds of sweet orange (*Citrus sinensis* [L.] Osbeck) fruits. Journal of Food Biochemistry. 2015;**40**(1):53-60

[47] Atti-santos AC, Rossato M, Serafini LA, Cassel E, Moyna P. Extraction of essential oils from lime by hydro-distillation and supercritical carbon dioxide. Brazilian Archives of Biology and Technology. 2005;**48**(1):155-160

[48] Mercy NP, Nithyalakshmi B, Aadhithiya LR. Extraction of orange oil by improved steam distillation and its characterization studies. International Journal of Engineering and Technology, Management and Applied Science. 2015;**3**(2):2349-4476

[49] Megha M, Mumtaj S. Extraction and characterization of essential oil of sweet lime (Citrus Limetta Risso) peel using microwave-assisted hydro-distillation. Research Journal of Chemical Sciences. 2014;**4**(11):51-55

[50] Ahmad MM, Rehman SU, Anjum FM, Bajwa EE. Comparative physical examination of various citrus peel essential oils. International Journal of Agricultural and Biology. 2006;**8**:186-190

[51] Jing L, Lei Z, Li L, Xie R, Xi W, Guan Y, et al. Antifungal activity of citrus essential oils. Journal of Agricultural and Food Chemistry. 2014;**62**(14):3011-3033

[52] Bocco A, Cuvelier ME, Richard H, Berset C. Antioxidant activity and phenolic composition of citrus peel and seed extracts. Journal of Agricultural and Food Chemistry. 1998;**46**(6):2123-2129

[53] Hertog MGL, Feskens EJM, Hollman PCH, Katan MB, Kromhout D. Dietary antioxidant flavonoids and risk of coronary heart disease: The Zutphen Elderly Study. The Lancet. 1993;**342**(8878):1007-1011

[54] Kawaii S, Tomono Y, Katase E, Ogawa K, Yano M. Quantitative study of flavonoids in leaves of citrus plants. Journal of Agricultural and Food Chemistry. 1999;**47**(9):3565-3571

[55] Garau MC, Simal S, Rosselló C, Femenia A. Effect of air-drying temperature on physicochemical properties of dietary fibre and antioxidant capacity of orange (*Citrus aurantium v. Canoneta*) by-products. Food Chemistry. 2007;**104**:1014-1024

[56] Khaoula Z, Amel BH, Ikbal C, Asma L, Jouda M-BJ. Chemical composition, fumigant and anti-acetylcholinesterase activity of the Tunisian *Citrus aurantium* L. essential oils. Journal of Industrial Crops and Products. 2015;**76**:121

[57] Olabinjo OO, Ogunlowo AS, Kamimura ES, Oliveira AL. Optimization of pressurized liquid extraction of essential oil from *Citrus sinesis* peels.

Agricultural Engineering International: CIGR Journal. 2020;**22**(2):255-263

[58] Vivian N, Giuseppe M, Bernardo R, Debora F. Citrus waste as feedstock for bio-based products recovery: Review on limonene case study and energy valorization. Journal of Bio-resource Technology. 2016;**214**:806-815

[59] Kim JW, Jo EH, Moon JE, Cha H, Chang MH, Cho HT, et al. In vitro and in vivo inhibitory effect of citrus Junos Tanaka peel extract against oxidative stress-induced apoptotic death of lung cells. Antioxidants. 2020;**9**:1231

[60] Anacleto SL, Milenkovic D, Kroon PA, Needs PW, Lajolo FM, Hassimotto NMA. Citrus flavanone metabolites protect pancreatic-beta cells under oxidative stress induced by cholesterol. Food Functions. 2020;**11**:8612-8624

[61] Huang R, Zhang Y, Shen SY, Zhi ZJ, Cheng H, Chen SG, et al. Antioxidant and pancreatic lipase inhibitory effects of flavonoids from different citrus peel extracts: An in vitro study. Food Chemistry. 2020;**326**:126785

[62] Rong X, Xu J, Jiang Y, Li F, Chen YL, Dou QP, et al. Citrus peel flavonoid nobiletin alleviates lipopolysaccharide-induced inflammation by activating IL-6/STAT3/FOXO3a-mediated autophagy. Food & Function. 2021;**12**:1305-1317

[63] Denaro M, Smeriglio A, Trombetta D. Antioxidant and anti-inflammatory activity of citrus flavanones mix and its stability after in vitro simulated digestion. Antioxidants. 2021;**10**:140

[64] Yang WL, Chen SY, Ho CY, Yen GC. Citrus flavonoids suppress IL-5 and ROS through distinct pathways in PMA/ionomycin-induced EL-4 cells. Food & Function. 2020;**11**:824-833

[65] Hu HJ, Zhang SS, Pan SY. Characterization of Citrus pectin oligosaccharides and their microbial metabolites as modulators of Immunometabolism on macrophages. Journal of Agricultural and Food Chemistry. 2021;**69**:8403-8414

[66] Smeriglio A, Alloisio S, Raimondo FM, Denaro M, Xiao JB, Cornara L, et al. Essential oil of Citrus lumia Risso: Phytochemical profile, antioxidant properties and activity on the central nervous system. Food and Chemical Toxicology. 2018;**119**:407-416

[67] Piccialli I, Tedeschi V, Caputo L, Amato G, De Martino L, De Feo V, et al. The antioxidant activity of limonene counteracts neurotoxicity triggered byAβ1-42 oligomers in primary cortical neurons. Antioxidants. 2021;**10**:937

[68] Cirmi S, Maugeri A, Lombardo GE, Russo C, Musumeci L, Gangemi S, et al. Flavonoid-rich extract of mandarin juice counteracts 6-OHDA-induced oxidative stress in SH-SY5Y cells and modulates Parkinson-related genes. Antioxidants. 2021;**10**:539

[69] Kitagawa T, Matsumoto T, Imahori D, Kobayashi M, Okayama M, Ohta T, et al. Limonoids isolated from the *Fortunella crassifolia* and the citrus junos with their cell death-inducing activity on adriamycin-treated cancer cell. Journal of Natural Medicines. 2021;**75**:998-1004

[70] Murthy KNC, Jayaprakasha GK, Safe S, Patil BS. Citrus limonoids induce apoptosis and inhibit the proliferation of pancreatic cancer cells. Food & Function. 2021;**12**:1111-1120

[71] Rekha CR, Vijayalakshmi G. Bioactive compounds in citrus fruits and

their health benefits. Food Science and Human Wellness. 2013;**2**(1):16-24

[72] Nazni P, Vijayalakshmi K, Sangeetha PT. Nutritional and functional properties of citrus peel flours. Journal of Agricultural Technology. 2011;7(2):607-619

[73] Wang Y, Chun OK, Song WO. Plasma and dietary antioxidant status as cardiovascular disease risk factors: A review of human studies. Nutrients. 2020;**12**(3):1-23

[74] Jayaprakasha GK, Patil BS. In vitro evaluation of the antioxidant activities in fruit extracts from citron and blood orange. Food Chemistry. 2007;**101**(1):410-418

[75] Vilas-Boas AA, Pintado M, Oliveira ALS. Natural bioactive compounds from food waste: Toxicity and safety concerns. Food. 2021;**10**(7):1564. DOI: 10.3390/foods10071564

[76] Raspo MA, Vignola MB, Andreatta AE, Juliani HR. Antioxidant and antimicrobial activities of citrus essential oils from Argentina and the United States. Food Bioscience. 2020;**36**:100651

[77] Gorinstein S, Cvikrová M, Machackova I, Haruenkit R, Park YS, Jung ST, et al. Characterization of antioxidant compounds in Jaffa sweeties and several citrus cultivars. Food Chemistry. 2016;**104**(4):1338-1344

[78] Dong W, Wei X, Zhang F, Hao J, Huang F, Zhang C, et al. A dual character of flavonoids in influenza a virus replication and spread through modulating cell-autonomous immunity by MAPK signaling pathways. Scientific Reports. 2014;**4**:7237

[79] De Clercq E. Potential antivirals and antiviral strategies against SARS coronavirus infections. Expert Review of Anti-Infective Therapy. 2006;**4**:291-302

[80] Tutunchi H, Naeini F, Ostadrahimi A, Hosseinzadeh-Attar MJ. Naringenin, a flavanone with antiviral and anti-inflammatory effects: A promising treatment strategy against COVID-19. Phytotherapy Research. 2020;**34**:3137-3147

[81] Zou Z, Xi W, Hu Y, Nie C, Zhou Z. Antioxidant activity of citrus fruits. Food Chemistry. 2016;**196**:885-896

[82] Ferreira SS, Silva AM, Nunes FM. *Citrus reticulata* Blanco peels as a source of antioxidant and anti-proliferative phenolic compounds. Industrial Crops and Products. 2018;**111**:141-148

[83] Kakkar S, Bais S. A review on protocatechuic acid and its pharmacological potential. ISRN Pharmacology. 2014;**2014**:952943

[84] Wu SC, Ng LT. Antioxidant and free radical scavenging activities of wild bitter melon (*Momordica charantia* Linn. var. abbreviata Ser.) in Taiwan. LWT – Food Science and Technology. 2018;**41**(2):323-330

[85] Silva S, Costa EM, Mendes M, Morais RM, Calhau C, Pintado ME. Antimicrobial and antibiofilm activity against *Staphylococcus aureus* of Opuntia flower extracts. International Journal of Molecular Sciences. 2014;**15**(1):21392-21402

[86] Cohen SM, Eisenbrand G, Fukushima S, Gooderham NJ, Guengerich FP, Hecht SS, et al. FEMA GRAS assessment of natural flavor complexes: Citrus-derived flavoring ingredients. Food and Chemical Toxicology. 2019;**124**:192-218

Chapter 4

Developing a Centralized Automatic Biomass Recovery System for Zero-Food Waste in High-Rise Apartment

Kyeong-Sik Kim, Dong-Hoon Lee, Yong-Woo Jeon, Soonkeum Han and Dong-Cheon Seo

Abstract

An advanced waste management system has been developed specifically for high-rise residential buildings, aiming to achieve zero-food waste emissions through efficient biomass recovery. In response to the global trend of increasing high-rise construction, this system includes two main components: a food waste cutting and transfer device installed under household sinks, and a centralized automatic module located in the building's basement. The system efficiently handles solid-liquid separation, dehydration, and bio-drying, converting food waste into high-quality biomass with low moisture content that is suitable for long-term storage. Fully automated and equipped with remote monitoring capabilities, the system ensures a sanitary and convenient environment for residents. Successfully implemented in two large-scale apartment complexes in Korea, it has operated seamlessly for over three years without causing any inconvenience to residents. Additionally, the system has encouraged greater resident participation in sustainable waste management practices, such as the separation of recyclable and special wastes in communal sorting rooms. The implementation of this integrated system is proposed as a viable technical solution for sustainable waste management in high-rise buildings, aligning with global trends in eco-friendly urban development and addressing the challenges posed by climate change.

Keywords: food waste, biomass recovery, bio-drying, zero-food waste, high-rise apartment

1. Introduction

Humans have built high-rise buildings for various reasons, and the height of buildings has various meanings. The centers of high-rise buildings, which began for religious reasons in ancient times, moved from North America to Asia and the Middle East in modern times. Currently, high-rise buildings and apartments exist not only in both large and regional cities [1]. In Korea, the supply of apartments and high-rise buildings is increasing dynamically, given economic development and social change. Owing to urban redevelopment and new city construction over the past 20 years, seeing skylines change owing to high-rise apartment complexes is not rare, not only in metropolitan areas but also in local cities. National housing statistics show that, among newly allocated homes, apartments increased from the 60s in 2010s to 82.0% in 2022 [2].

Meanwhile, the problems associated with the management of solid waste in today's society are complex because of the quantity and diverse nature of waste, the development of sprawling urban areas, funding limitations for public services in many large cities, the impacts of technology, and emerging limitations in both energy and materials. Moreover, households are the subject of the second of the six functional elements in the solid waste management system, which includes waste handling and separation, storage, and processing at the source. Waste handling and separation involves the management of wastes until they are placed in storage containers for collection and also encompasses the movement of loaded containers to the point of collection, which is an important step in the handling and storage of solid waste at the source [3]. Therefore, the second functional element of waste management is closely related to housing structures and waste collection systems.

In Korea, a mandatory food waste separation system enacted 20 years ago, motivated by a volume-based waste fee system enacted 30 years ago, has had a groundbreaking impact on household waste management. Among separable household waste items, food waste is the most difficult to separate. Therefore, appropriate measures must be established at the household, local government, and national levels in accordance with changes in the economy, society, and the environment. Although introducing an innovative mandatory separate collection system has quickly brought the country's waste management level beyond that of developing countries, methods of responding to food waste according to the path appropriate for the continued improvement of living standards and social change must also change. Considering the chronic problems of food waste from the source to the recycling market, as well as the characteristics of a high-rise apartment society, an integrated solution using a zero-food-waste system through centralized and automatic biomass recovery with mechanical-biological treatment (MBT) and information and communication technology (ICT) convergence technology is proposed for next sustainable high-rise apartments. The number of people who want to live in high-rise apartments with a pleasant view is increasing, but none want to separate and store food waste at home or meet neighbors carrying bags of food waste in closed hallways or elevators. Zero-food waste emissions may be the only solution.

2. Impacts of food waste and separate food waste collection system

The world generates 0.74 kg of waste per capita-day, yet national waste generation rates fluctuate greatly from 0.11 to 4.54 kg per capita-day, and waste generation volumes generally correlate with income levels and urbanization rates. The largest waste category is food and green waste, accounting for 44% of global waste at the international level, more than 50% of the waste in low- and middle-income countries, and 32% of the waste in high-income countries. However, the amount of organic waste is widely comparable in absolute weight terms because of the larger amounts of packaging and because of other categories such as income levels. Globally, more than

95% of lower-income countries, more than 80% of middle-income countries, and approximately 40% of waste are disposed off in open dumping and landfills. In other words, the rates of open dumping and landfilling decrease as a country's income level increases [4].

The U.S. EPA investigated the environmental impacts and contributions to a circular economy of 11 common pathways to manage wasted food, from source reduction to composting to landfill. In its report, the term 'wasted food' was used to encompass both excess food and food waste, and two methodologies—life cycle assessment (LCA) and circularity assessment—were employed to evaluate these pathways. Scale emphasizes the importance of preventing and diverting food waste from sewer/wastewater treatment, landfills, and controlled combustion (i.e., incineration) pathways. In addition to its impact on the environment, food waste has negative social and economic impacts; therefore, integrated measures are required [5].

Governments in increasingly more countries and localities are promoting separate events to reduce food waste or processing and recycling byproducts through separate collection. When food waste is reduced at its source through separate collection, the quality of the remaining household waste improves, leading to easier hygiene management and resource recovery.

In 2023, New York City implemented the largest composting program in the United States through a curbside composting program with a simple, universal weekly collection of leaves, yard waste, food scraps, and food-contaminated paper products for cleanliness and sustainability [6]. San Francisco launched the nation's first and largest municipal food waste composting collection program in 2002, covering both commercial and residential sectors and setting the city as a model green city [4].

The pipeline collection of ground food waste from households to biogas plants has been performed by local governments in Sweden and other Nordic countries [7]. In Burkina Faso, decentralized organic waste management across households can considerably reduce the burden on formal waste collection and disposal infrastructure while improving food security through agricultural benefits and creating opportunities for citizens to generate income from waste [4].

Two systems of volume-rate waste fees for household waste and separate collection of food waste have become motivations for reforming the concept of municipal solid waste management in Korea from open dumping and burning to sustainable waste management policies. Approximately 60% of household waste is separated for recycling, and over 90% of separately collected food waste is transferred to facilities for feedstock of animal feed, compost, and biogas [8]. Separate collection of food waste has made residual household waste easier and more beneficial for physical separation and thermal treatment. However, owing to rapid changes in global issues and the social style of living, this study aimed to report on technology developed for integrated measures through zero-food waste for the social changes of high-rise apartments and a carbon reduction society.

3. Pilot study of automatic solid recovery system for food waste

3.1 Purpose and scope of pilot study

This study assessed a drainage pretreatment system equipped with a solid-liquid separation system and dehydrator, and 240 of the 283 apartment households (participation rate of 88%) participated in three pre-set installation sites. The purpose of this pilot study is as follows, and the connection structure with the sewerage facility is shown in **Figure 1**.

1. Resolve citizen inconvenience caused by the unhygienic separate collection of food waste and introduce a food waste source treatment and resource recovery system with a treatment method suitable for various residential environments.

2. By using the plumbing facilities of existing apartments (apartments and villas), kitchen sewage with crushed food waste is separated into solids and liquids, which are recovered and used as resources. The sewage is then discharged through a filter into a sewer pipe.

3. Review cutting and transfer devices, solid-liquid separators, and resource recovery devices, among others, and obtain basic data for establishing institutional standards, such as deriving design data.

3.2 Materials and methods

The system of pilot study consisted of a cutting and transfer device, solid-liquid separator, resource recovery device (dehydrator), and pluming. Monitoring of the system included the identification of food waste generation characteristics before and after system installation, kitchen sewage discharge characteristics, performance, possibility of recycling recovered solids, resident awareness, and adaptability to indoor piping and public sewers.

As for a basic piping plan, the system was connected to an existing indoor horizontal branch pipe, underground horizontal pipe, and kitchen sewage vertical pipe, in accordance with the standard specifications and building codes recommended by the Ministry of Land, Transport and Maritime Affairs. Transparent observation pipes and cleaning ports were installed in the transverse pipe for obstruction observation and cleaning.

The cutting and transfer device had a processing capacity of 1.0 L/time and consisted of a main body, automatic tap water supply, and control switch. A shredding method was a combined cutter and hammer mill with a discharge pressure of 1.5 m-H_2O by rotating a multi-stage impeller to prevent blockage of the transverse branch pipe. Automatic and manual operation types existed, and the safety of each device was considered in its design.

The solid-liquid separator had a repeated stroke of a pair of filters linked to a water level sensor and a controller based on environmental settings. A solid slurry transfer pump and screw press are operated in conjunction, and a discharge pump

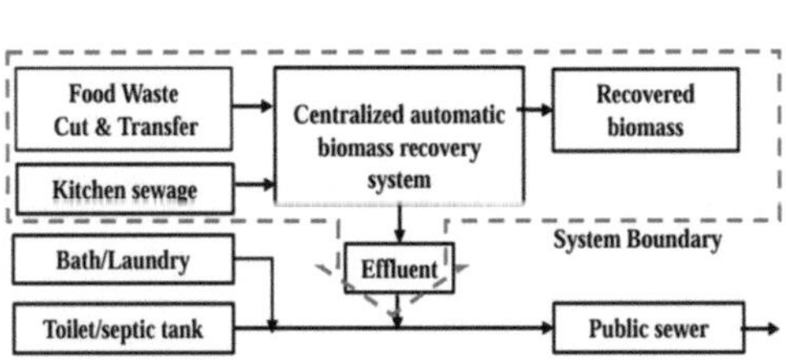

Figure 1.
Concept of automatic biomass recovery system and public sewer in pilot study.

and makeup water circulation pump are operated using a water level sensor. The self-diagnosis functions were displayed on an LCD monitor. It consisted of a submersible solid transfer pump and a pair of filters equipped with submersible transfer pumps. Filter regeneration uses the centrifugal force generated by rotating the filter at high speed to separate and regenerate solids attached to the filter medium. The water level sensor detected the regeneration cycle according to the flow rates of the inflow and discharge water. In addition, dissolved oxygen was generated via aeration during regeneration, creating aerobic conditions that did not produce odors under anaerobic conditions. Dual solid-liquid separation filters were installed such that, even if one filter failed, the other filters operated normally, and the filtration function was always maintained. The treatment capacity was set to less than 100 kg/day in accordance with Annex 3 of the Enforcement Decree, Waste Management Act of Korea [9].

The resource recovery device was a screw dehydration-type device equipped with various functions, such as self-failure diagnosis, environmental settings, freeze prevention, temperature and humidity control, and weight integration. Dual units for safety were installed in the event of a malfunction. These consisted of two screw presses, an electronic scale, an LCD monitor, a system controller, and a recovered solid material container.

System monitoring was conducted for 10 months before and after system installation, and at least three seasons (winter, spring, and summer) were evaluated to determine seasonal suitability. Before installing the system, a survey was conducted to investigate the basic characteristics of kitchen sewage. BOD_5, SS, pH, n-hexane extractable material, T-N, and T-P were analyzed to determine the water quality of the kitchen sewage.

After installing the system, the characteristics of the discharged sewage, the possibility of recycling the recovered biomass, system performance, residents' awareness (based on a survey), and adaptability to indoor piping and public sewage were investigated. Key evaluations were conducted according to the certification standards of public certification bodies. The process and pictures of the prototype biomass recovery system installed in the three apartments for the pilot study are shown in **Figure 2**.

3.3 Summary of results and discussion

A pilot study on the food waste reduction effect of a centralized resource recovery system was performed by monitoring for 6 months after system installation. The results and discussion are summarized as follows.

1. By reviewing the survey results and reference data, it was found that 30 L/capita-day of the unit generation amount of kitchen sewage was reasonable before system installation, and 5 L/capita-day of the unit's added amount after system installation was also reasonable, which is the same as the recommended water amount in the user manual of the cutting and transfer device. When considering the total daily amount of sewage generation, which is 441 L/capita-day in Seoul [10], the amount of 5 L/capita-day added by cutting and transfer devices is negligible.

2. The water qualities of the effluents before (n = 8)/after (n = 9) system installation were 335/283 mg/L for BOD_5, 221/113 mg/L for SS, and 113/44 mg/L for n-hexane extractable material in the Y apartment. Moreover, the following values were obtained in the final test after the system installation: 256.3 mg/L for BOD_5, 142.5 mg/L for SS, and 91.4 mg/L for n-hexane extractable material. Both

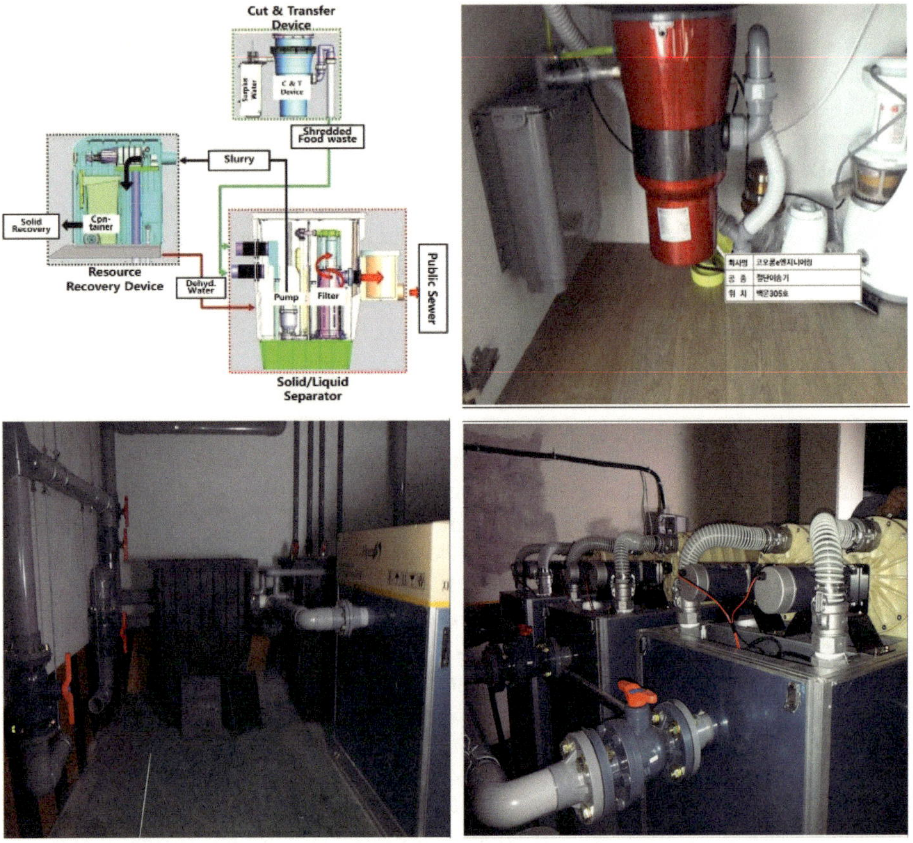

Figure 2.
Prototype biomass recovery system installed in pilot study.

cases were within the medium-to-strong concentration ranges of typical sanitary wastewater [11] and became lower after system installation. For reference, a single device performance test was conducted to evaluate the basic performance of the system under simulated food waste and test conditions (food waste discharge of 200 g/person-day, kitchen sewage amount of 35 L/person-day; 500 persons). The results of effluent water quality were 166.2 mg/L for BOD_5/L and 170.6 mg/L for SS, which were lower than the pilot study. That is, contrary to concerns, it was confirmed that the centralized automatic biomass recovery system could somewhat improve the effluent water quality in the pilot study.

3. A survey on the food waste reduction effect before and after system installation in the pilot study was performed using measured data for food waste collected after system installation and statistical data from the year before system installation. The results showed a 34.1% reduction in the first month after system installation, a 55.3% reduction in the second month, and an average of 70% reduction from the third to the fifth month. Considering that 84.0% of the households participated in the pilot study, it was estimated that the waste reduction effect due to system installation reached a maximum of 84.0%, even in the first pilot study. Therefore, there was a highly positive effect on achievable the 100% reduction of food waste in further steps.

4. Mass balance analysis was conducted on the solid recovery system in this pilot study to calculate the amount of total solids (TS) discharged into the sewer and the solid biomass recovered from the system. The solid recovery rate was calculated to be 54.6% in the first analysis 3 months after system installation, which increased to 78.7% in the fifth analysis after 5 months, and the TS discharged into the sewer showed a maximum of 46.7% and a minimum of 9.1%. Because biomass recovery and discharge rates vary depending on the composition of food waste and the sampling time, long-term monitoring is necessary to obtain reliable data. Photographs of the biomass recovered during the pilot study are shown in **Figure 3**. The moisture content was in the range of 65–70%. The salt content was 0.18–0.09%, which was within the legal standards of 2.0% set in the fertilizer process standards [12], 0.4–0.5% for pig feed, and 0.15–0.37% for chicken feed [13].

5. An awareness survey of residents was conducted twice before and after installation regarding the cutting and transporting device installed in each household's kitchen. The overall response rate was 51% before installation and 84% after installation, which indicates that interest increased after installation. Respondents were women (70%), people over forties (83%), and households over three people (82%). More participation from residents who had higher needs for it were answered in the survey. Among all types of waste subject to separate collection, the most inconvenient waste is food waste (70%), followed by bulky waste (19%), residual waste subject to volume-rate system (7%), and recyclable waste (4%). The most inconvenient aspect of the food waste separate collection system was the discomfort and hygiene issues when handling for sorting and storing food waste (59%), followed by the process of carrying it by hand to the collection point (17%). Respondents who separate and discharge food waste every day are 23%, every 2 days, 28%, and every 3 days, 32%. In other words, 84% of respondents were experiencing the inconvenience of regularly separating and discharging food waste at intervals of less than 3 days. Regarding this, 87% of respondents answered about the convenience of using the cutting and transfer device installed in each household's kitchen, but some appealed to the inconveniences caused by lack of experience with the machine. When respondents were asked about increases in utility rates, 83% of respondents said they had not noticed increases in electricity and water rates.

6. As a result of considering the feasibility of introducing the system along with a mechanical evaluation to improve the system performance, this system can be an

Figure 3.
Recovered biomass samples in the pilot study.

alternative that can solve the problem of convenience for residents regarding a separate collection system of food waste. It can also be a fundamental solution to the problems of food wastewater on land treatment due to the lack of problems with system operation, such as sewer pipe sediment, noise, and odor. A win-win effect may be expected by utilizing the advantages of the automatic biomass recovery system within the limit of not harming the sewage environment and also taking advantage of the carbon reduction effect due to the increase of biomass circularity and zero-food waste emissions. As for the food waste problem, it could be possible to build a sustainable food waste management for high-rise apartment complexes.

4. Development of full-scale zero-food waste system in high-rise apartments

Various policies to save food and reduce food waste have had many positive effects on waste management. However, to solve the appropriate chronic food waste problems, as well as economic and social changes, technological development is required. Therefore, not only the source reduction but also the waste disposal stage, optimal sustainable waste management with zero landfills and zero incineration through innovative technology is required. In this section, the development of technology that can be applied to sustainable high-rise apartments by solving chronic problems in the separation and recycling stages of food waste sources [14] by improving the pilot study results is described.

4.1 Goals and tasks

The most important aim of this study is to achieve zero-food waste through a centralized automatic biomass recovery system in high-rise apartments. Therefore, a full-scale zero-waste system is another title for this system. This system is to provide a convenient, hygienic, and eco-friendly solution to chronic problems and carbon reduction effects by zero-food waste emissions for the ever-increasing high-rise apartment society. The recovery of high-quality biomass in a ready-to-use state can reduce the burden on local governments by eliminating the need for separate collection, transportation vehicles, and recycling facilities. To further complete the full scale system, the objectives of the pilot study are as follows:

1. Improvement of hygienic, clean, and convenient system

2. Improvement of technical system with reliability

3. Zero-food waste by biomass recovery in high-rise apartment.

4.2 Improvement of full-scale system using MBT-ICT convergence technology

The pilot study yielded many important findings, including the suitability of the system functionality, potential for zero-food waste emissions from solid resource recovery, minimal impact on the sewage environment, good quality of recovered biomass resources, and minimal impact on utility consumption. However, the prototype system of the pilot study was not suitable for application in a full-scale system of high-rise apartments or for marketing the technology. Therefore, functional, structural, and performance improvements were necessary. The full-scale system also consisted of cutting and transfer, centralized automatic biomass recovery, and piping facilities connecting them. Accordingly, the basic concept of the system was reimagined and redesigned as ergonomic and eco-friendly concepts by adding appropriate functions and rearranging the process of MBT and ICT convergence technologies. A bio-drying process using biological treatment technology was integrated with MT to reduce the moisture content of the recovered biomass.

The moisture content of the biomass recovered in the pilot study was 65–70% in the dehydrated state, and decomposition by microorganisms may cause biomass deterioration and odor. Therefore, drying was necessary to reduce the moisture content as much as possible to a level where microorganisms cannot function. Because a thermal drying method is not suitable for developing this system because of the burden of the external energy supply, a bio-drying process that does not require external energy rather uses the heat of decomposition of the biomass itself to dry it was applied [15, 16].

As a result, the temperature within the process increased to approximately 40–50°C, allowing the heat necessary for drying to be supplied, and the moisture content of recovered biomass could be reduced to less than 15% by supplying excess air. This mechanism could be explained by the fact that evaporation was driven by the difference between the saturated water vapor pressure of the waste surface and the water vapor pressure in the passing air, which increased either due to temperature increases associated with biodegradation or the dryness of the introduced air. A high airflow rate (AFR) can enhance the water removal rate, and water removal associated with air replacement was generally greater than that associated with temperature increases caused by biodegradation. However, an excessive AFR terminated biodegradation owing to the reduced moisture content, even though organics remained [15]. The AFR was confirmed as a key operational parameter significantly affecting both the bio-drying performance and efficiency [16]. Excess air created an aerobic atmosphere inside the module, thereby preventing anaerobic decomposition.

Additionally, the automatic module containing these three processes was redesigned from a horizontal structure in the pilot study to an upright, human-like vertical structure in the full-scale system. Consequently, materials within the system could be moved by gravity, and the top-down upright system structure can save space and energy. For automation and unmanned management of the entire system, interlocking and ICT functions were added, and compact modularization can help with installation and maintenance convenience.

4.3 Installation and operation of full-scale system for high-rise apartment

The zero-food waste system for high-rise apartments using MBT-ICT convergence technology consisted of cutting and transport devices, dual automatic modules, and single manual modules in a dedicated space on a basement floor, and it was installed in each building. This system was installed as a highly complete built-in unit in about 1451 households in nine buildings of 35 floors in Seoul, South Korea,

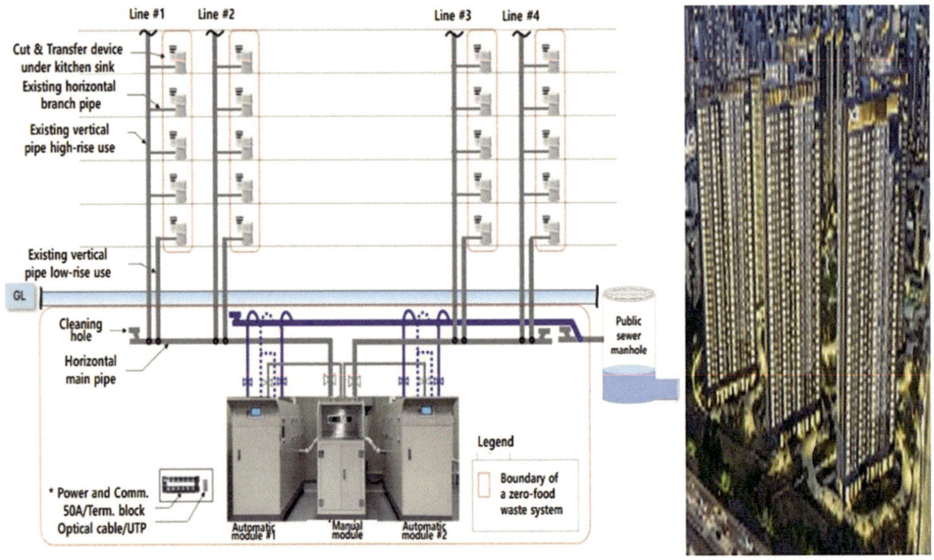

Figure 4.
Layout of a zero-food-waste system for high-rise apartments using MBT-ICT convergence technology.

after a three-month construction period starting in July 2021, and it has been operating without any breakdown for 3 years. The layout of the system is shown in **Figure 4**.

To prepare for system operational failures and special situations, all buildings were equipped with multiple modules (a pair of automatic modules and one manual module). The two automatic modules can operate simultaneously or independently, and a manual module is implemented to prepare for emergencies. Each automatic module was equipped with a solid-liquid separation process, dewatering process, bio-drying process, and temporary storage box of recovered biomass in a vertical structure (see **Figure 5**). Each process was connected by an interlocking device such that each process was automatically operated individually or simultaneously. Signals regarding the operating status of the parts were displayed on the manager's cell phone so that remote managers and related personnel could check them at any location, eliminating the need for a resident manager.

The food waste placed into the cutting and transfer device of each household is cut and reduced to an appropriate size within 30 s by a multi-stage impeller and transferred to an underground automatic module through the horizontal and vertical kitchen sewage pipes. This system operates automatically and intermittently, when necessary, 24 hours a day, and even if one module breaks down, the other modules start operating. In the case of a disrupted process, the failure of a specific part is displayed on the LCD, and notification is sent concurrently with the manager's smartphone so that repairs can proceed quickly. An emergency manual module can be operated in the presence of a manager under abnormal conditions, such as when excessive food waste is generated or when the cutting and transfer device in homes is out of order. This means that the system has a dual emergency measure by the monitoring and emergency call line connected to an automatic remote notification to relevant personnel and the integrated management server of the central disaster prevention center.

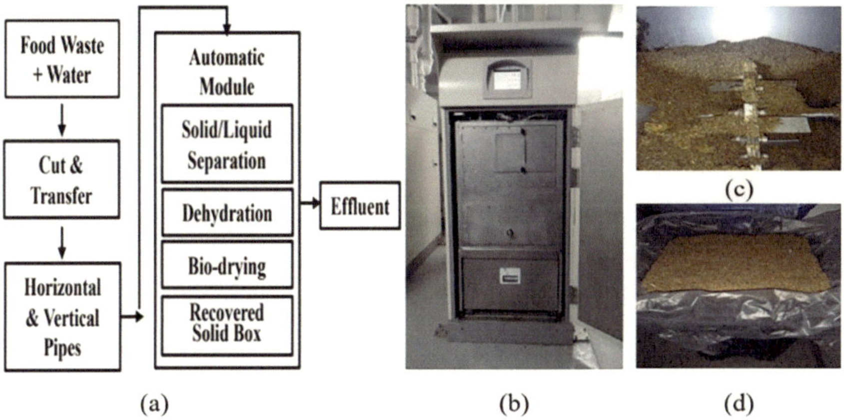

Figure 5.
Flow chart (a) and picture of automatic module inside (b), and bio-drying process (c), and bio-dried biomass in temporary storing box (d).

4.4 Results and discussion

Solid waste management may be defined as the discipline associated with the control of generation, storage, collection, transfer and transport, processing, and disposal of solid wastes in a manner that is in accord with the best principles of public health, economics, engineering, conservation, esthetics, and other environmental considerations, and that is also responsive to public attitudes. In its scope, solid waste management includes all administrative, financial, legal, planning, and engineering functions involved in solutions to all problems of solid waste. The solution may involve complex interdisciplinary relationships among such fields as political science, city and regional planning, geography, economics, public health, sociology, demography, communications, and conservation, as well as engineering and material science [17]. Integrated solid waste management (ISWM) can be defined as the selection and application of suitable techniques, technologies, and management programs to achieve specific waste management objectives and goals [18].

In order to solve the food waste problem, complex relationships in all processes of the food waste management system must be considered, but in order to achieve specific purpose waste management, it is necessary to select and apply integrated and appropriate technologies. This study was conducted to propose a technically integrated solution against chronic problems of the present food waste management system in the manner of best principles and the public attitude.

4.4.1 Improvement of hygienic, clean, and convenient system

The first purpose of the pilot study sponsored by the Seoul Metropolitan Government is "to resolve citizen inconvenience caused by the unhygienic separate collection of food waste and introduce a food waste source treatment and resource recovery system with a treatment method suitable for various residential environments." By improving the results obtained in the pilot study, the full-scale project of a zero-food waste system for high-rise apartments achieved its goal in accordance with the best principles and public awareness and in accordance with the increasing trend of high-rise apartments.

This system is a convenient system that disposes of food waste at the sink in a household without having to take it out of the home. That is, the path through which food waste passes is a closed sewer pipe rather than a hand-carried elevator, and the centralized automatic biomass recovery system located underground is also packaged in compact style. Aerobic condition in the module is maintained by fresh excess air of the bio-drying process. Therefore, odor generation conditions do not exist inside the module.

As a result of operation and maintenance over 3 years, it was confirmed that the hygienic, no-odor, convenient, and zero-food waste system could be provided to households by the recovery of high-quality biomass. In addition, the recovered biomass can be recycled in fresh condition for a long period, allowing it to be directly recycled as compost and biomass solid fuel without additional processing with long-term storage. **Figure 6** shows the recovered biomass that can be stored long without changing the quality at room temperature.

4.4.2 Improvement of MBT-ICT system: Effluent quality

Effluent which is mixed water of water from the solid-liquid separation device and dehydrated water from the dehydrator is discharged into the public sewer. Although there are no effluent water quality standards in Korea, sediment should not deposit inside the sewer and the effluent from the system should not be impacted on the sewage treatment plant.

Figure 7 shows the daily changes in water quality of effluent randomly sampled at intervals of 1–2 h except late at night on November 7, 2023 (Tuesday) and May 19, 2024 (Sunday). The daily average values of the effluent water quality for both days were near the typical medium sanitary wastewater concentration in the United States [11] and were lower except for n-hexane extractable material than the effluent standard of the drainage treatment type of disposer system in Japan, which are 300 mg/L in BOD_5, 300 mg/L in SS, and 30 mg/L in n-hexane extractable material [19]. In other words, it is believed that the development of the full-scale system has increased the reliability of the discharge water quality of the full-scale system compared to the pilot study. In the case of n-hexane extractable material, it is considered to be somewhat high as there is no such standard in Korea. There was no case of shock load to a sewer system, and even in cases where it was somewhat high, the impact on the sewer system is expected to be minimal because the amount of kitchen sewage is less than 1/10 of the total amount of domestic sewage. That is, it is believed that the

Figure 6.
Photos of recovered biomass stored at room temperature for 280 days, 140 days, 70 days, and 1 day since September 6, 2023.

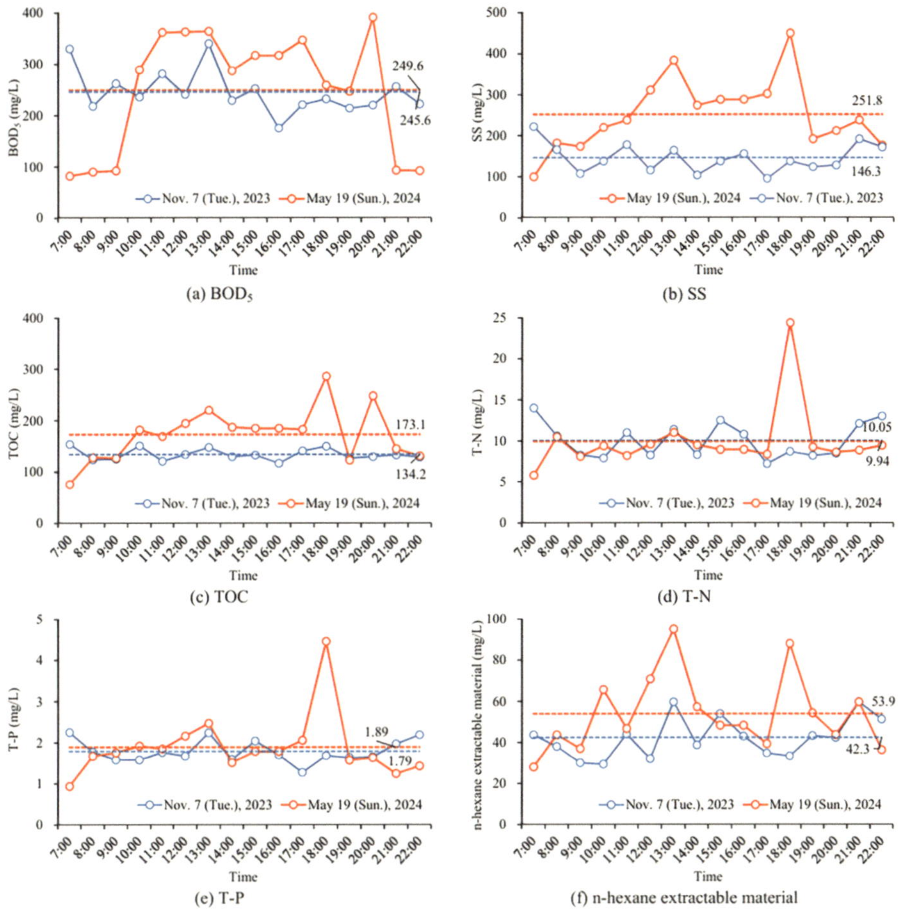

Figure 7.
Daily variation and average of water quality of effluent from the full-scale system on Tuesday, November 7, 2023, and Sunday, May 19, 2024.

development of the full-scale system has increased the reliability of the effluent water quality compared to the pilot study.

4.4.3 Improvement of MBT-ICT system: Recovered biomass

Table 1 shows the results of the analysis of the compost and feed quality of the biomass recovered from food waste by this zero-food-waste system. The quality of the compost and feed met legal standards. In particular, many heavy metals were undetected or were detected at very low levels, whereas *Escherichia coli* and *Salmonella* were not detected. The organic-matter–nitrogen ratio, moisture content, and salt content were also very low, and all met the legal standards. This was possible because the food waste discharged from homes was collected in a very fresh state, washed with water, and dried to a moisture content of less than 15% via the bio-drying process. Therefore, the recovered biomass did not deteriorate biologically, even when stored for a long period.

Table 2 shows an analysis of the fuel characteristics. The system produces an excellent quality bio-SRF and can be stored for long periods of time. Additionally,

Compost quality				Feed quality			
Item	**Unit**	**Result**	**Legal Stand.**	**Item**	**Unit**	**Result**	**Legal Stand.**
OM	%	82.75	≧30	Moisture	%	13.70	≦14
Arsenic	mg/kg-dry	N.D.	≦45	Crude protein	%	22.79	—
Cadmium	mg/kg-dry	N.D.	≦5	Crude fat	%	3.84	—
Mercury	mg/kg-dry	N.D.	≦2	Crude fiber	%	24.14	—
Lead	mg/kg-dry	0.68	≦130	Crude ash	%	3.22	—
Chrome	mg/kg-dry	6.03	≦200	Salinity	%	0.29	—
Copper	mg/kg-dry	12.29	≦360	Phosphorus	%	0.20	—
Nickel	mg/kg-dry	2.66	≦45	Lead	ppm	0.01	≦20
Zinc	mg/kg-dry	39.81	≦900	Mercury	ppm	≦0.01	—
E. coli	—	N.D.	N.D.	Cadmium	ppm	N.D.	—
Salmonella	—	N.D.	N.D.	Arsenic	ppm	0.01	—
ON	%-	23.71	≦45	Fluorine	ppm	N.D.	—
Salinity	%	0.13	≦2.0	Copper	ppm	10.52	≦29
Moisture	%	13.98	≦55	Zinc	ppm	20.58	≦87
Maturity (S)	—	4	≧4	Selenium	ppm	0.09	≦2
HCl insol.	%	0.42	≦25	Aflatoxin	ppb	N.D.	≦50
Nitrogen	%	3.49	—	Salmonella	—	N.D.	N.D.
pH	—	5.15	—	Melamine	—	N.D.	—

OM, organic matter; E. coli, E. coli O157:H7; ON, organic nitrogen; Maturity (S), maturity (Solvita).

Table 1.
Qualities and standards for compost and feed of recovered biomass.

Item		Unit	Result
Proximate analysis	Moisture	%	14.06
	Volatile matter	%	67.06
	Fixed carbon	%	16.15
	Ash	%	2.74
Heating value	HHV (dry basis)	kcal/kg	4843
	LHV (wet basis)	kcal/kg	3759
Ultimate analysis	C (carbon)	%	48.85
	H (hydrogen)	%	6.87
	O (oxygen)	%	40.58
	N (nitrogen)	%	4.03
	S (sulfur)	%	0.25

Table 2.
Quality analysis results for bio-SRF of recovered biomass.

the biological methane potential (BMP) was approximately 0.339 Nm^3 CH_4/kg VS or higher under the substrate-to-inoculum (S/I) ratio of 0.3, meaning that a considerable amount of biodegradable organic substances remained in the biomass. Meanwhile, when calculating the theoretical amount of methane generated according to the Buswell and Mueller equation, it was found to be 0.477 Nm^3 CH_4/kg VS, which is similar to that of food waste, supporting the BMP test results [20, 21]. Considering the above results, zero-food waste emissions can be achieved by producing excellent biomass resources that can be used directly from food waste without a separate treatment system.

To verify the quantitative value, evaluating the amount of resources that could be secured is necessary. However, this is difficult to estimate immediately because there were not enough cases. In this study, a simple comparison between energy usage and existing food waste recycling methods was performed.

4.4.4 Comparison of energy usage with existing methods

Waste is a by-product of economic activities, and sustainable solutions based on market economic principles must be sought for a circular economy. However, because waste management policy in Korea is based on government subsidies rather than market economic principles, decision-making in the market is difficult. For such decision-making, methods such as LCA, life cycle cost, social LCA, and Life Cycle Sustainability Assessment (LCSA) have been proposed; however, these methods have limitations in terms of standardization and application owing to the different situations of the three pillars of economy, society, and the environment. Therefore, instead of evaluating the impact of the entire process, this study only compared the energy consumption of food waste disposal methods.

The comparison target was one building (174 households) among the high-rise apartment complexes where this research system was installed. Based on the same criteria, the centralized automatic biomass recovery system using MBT-ICT convergence technology (hereinafter zero-food-waste system) was compared with feeding and composting system, which are representative food waste treatment technologies

Category	Basic operating conditions
Persons	174 × 2.41 person/household [22]
Emission unit (apartment)	209.72 g/capita-day [23]
Number of times food waste is discharged (S1, S2)	2.02 times/week [24]
Energy consumption per elevator use (S1, S2)	30 Wh [25]
Amount of kitchen sewage generated (S3)	30 L/capita-day [26]
Water for device (S3)	5 L/capita-day [26]
Power consumption (S3)	15.89 kWh/day (measured value)

Table 3.
System operating conditions.

that have been used for a long time. Bio-gasification not only integrates the treatment of organic waste such as sewage, food waste, and excrement but also uses the produced biogas as a heat source for facilities, so it was excluded from comparison due to different standards for energy usage. **Table 3** shows the operating conditions of the technologies being compared. For simplicity, dry feeding is indicated as (S1), composting as (S2), and zero-food-waste system as (S3).

Table 4 shows the pathway of food waste for this comparative analysis research. Food waste has high moisture and biodegradable contents. Therefore, it is easily perishable at room temperature, and a large amount of leachate is produced physically and biologically during every stage of pathways. The leachate is difficult to treat at every stage and is sent to a nearby sewer or a separate treatment facility, but most of it is sent to a sewage treatment plant. The impact on sewage caused by leachate is the same in all cases, so it can be ignored in comparison. Here, the value of recycled products by the food waste recycling process cannot be judged normal due to chronic problems in the recycling market [14], so it was not included in the calculation.

When the total energy consumption required to process 1 ton of food waste in each process was converted into tons of oil equivalent (toe), as shown in **Table 5**. The dry feeding process (S1) showed the largest energy consumption; the next was the zero-food waste system (S3) of this study, and the last was the composting process (S2). Although the difference between these three processes was not so big, the zero-food waste system (S3) consumed 14.4% less than dry feeding (S1) but 26.7% more than composting (S2). But the dry feeding (S1) consumed 51.1% more energy than composting (S2). Although no meaningful calculations have been made based solely on energy usage by process, the zero-food waste system (S3) provided an approximately

System	Step
Dry feeding (S1)	Discharge (by elevators) → Collection → Transportation → Dry feeding ↘leachate ↘leachate ↘leachate ↘leachate to sewer
Composting (S2)	Discharge (by elevators) → Collection → Transportation → Composting ↘leachate ↘leachate ↘leachate ↘leachate to sewer
Zero-food waste (S3)	Cutting and transfer device → Centralized automatic biomass recovery system ↘Effluent to sewer

Table 4.
Food waste processing flows for this comparative analysis research.

Scenario	Sum	Step		
		Separation	Collection-transportation	Disposal
S1	0.04733	0.00392	0.00331 [27]	0.04010 [28]
S2	0.03033			0.02310 [28]
S3	0.04138		0.04138	

Table 5.
Energy use by scenario. (Unit: toe/t-FW).

570 times reduction of elevator uses compared to composting (S2) and dry feeding (S1) when processing 1 ton of food waste. That is, these results confirmed that the zero-food waste system (S3) of this study provide not only the biggest advantages of zero-foot waste emission through biomass recovery but also convenience and hygienic environment to residents.

4.4.5 Synergy effect on resident's voluntary participation in sustainable behaviors of waste management

Northern European countries, such as Finland and Denmark, have high global happiness scores, according to the annual announcement by the United Nations [29]. Interestingly, sustainable behaviors regarding waste management were linked to sustainable happiness because happier people are more likely to engage in waste management and sustainable behaviors, and waste management and sustainable behaviors induce happiness in people [30].

In high-rise apartment buildings, where full-scale centralized automatic biomass recovery systems have been installed, a communal waste separating room has been operated for the voluntary separation of recyclables and special items in waste. It is located between two elevator entrances and connected to an underground communal parking lot. **Figure 8** shows the residents' communal waste separating room being operated cleanly and a centralized automatic biomass recovery system room (usually locked). In the communal separating room, various types of household waste, excluding food waste, were easily collected through the voluntary participation of residents so that the waste could be sorted by item for recycling or appropriate disposal. The contents shown in each photograph are as follows: (a) food waste facility room (usually locked), auxiliary container for volume-rate fee waste, waste home appliances, waste furniture, waste fluorescent lamps, and waste batteries; (b) waste cooking oil, emergency food waste, non-deposited glass bottles, deposited glass bottles, ceramics, disposable ice packs, handbags, shoes, carpets, curtains, blankets, and used clothes; (c) scrap metal, cans, vinyl, transparent plastics, Styrofoam; (d) cardboard, other plastics, and three types of papers; (e) an communal waste separating room between the underground elevators and parking lot (f) a communal waste separating room of next building and food waste facility room (open for inspection). This voluntary participation in waste separating behaviors for 24 items is possible because of no hand sorting for food waste included to residents. Ensuring zero-food waste emission through a centralized automatic biomass recovery system can have a synergic effect on sustainable behaviors, which is participation in waste separation practices in this case. Therefore, a sustainable high-rise apartment may be created by a higher resource circulation rate. An increase of voluntary participation in sustainable behaviors may be linked to sustainable happiness and sustainable high-rise apartments and cities further.

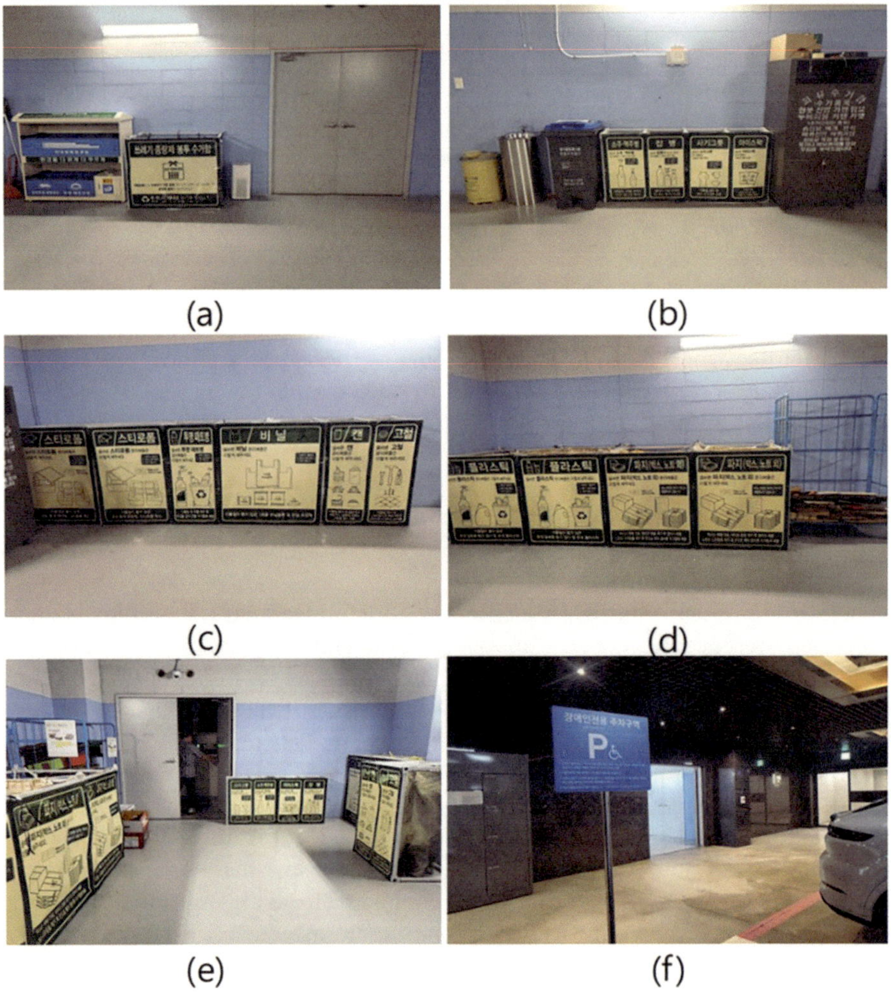

Figure 8.
Photos of residents' communal separation room (24 items) for recyclable and special items and food waste treatment facility room since July 2021.

5. Conclusions

The supply of apartments and high-rise buildings is increasing dynamically, given economic development and social change in Korea. An integrated system appropriate for high-rise apartments was developed to eliminate insanitation and inconvenience during the separation of food waste and to achieve zero-food waste emissions by recovering biomass as a ready-to-use resource. This system consisted of two parts: a cutting and transfer device of food waste under the sink of each household and a centralized automatic module in the basement. The cutting and transfer device has the shape of a typical disposer but different roles and functions owing to the special design. The centralized automatic module is a system in which three processes, namely solid-liquid separation, dehydration, and bio-drying, are combined with ergonomically and eco-friendly designed MBT-ICT. Therefore, a convenient and sanitary environment is created inside the apartment building. The effluent was in the typical range of medium sanitary wastewater concentrations, which could not cause troubles in a sewer system. And the recovered biomass had good quality and low moisture content, which could be recycled directly and long stored without further treatment. All the processes in the system are operated automatically, interlocked, and connected to the cell phones of remote and control center managers. This technology was developed for high-rise apartments by improving the results of a pilot study supported by the Seoul Metropolitan Government and redesigned for sustainable waste management. A full-scale system has been installed in two high-rise apartment complexes including 1451 homes in nine buildings with 35 floors without troubling residents for over 3 years. A technical solution approach to sustainable waste management for high-rise apartments was proposed by implementing a zero-food-waste system in this paper. In addition, synergistic effects on residents' participation in sustainable behaviors in waste management are expected with this zero-food waste system.

Author details

Kyeong-Sik Kim[1], Dong-Hoon Lee[2*], Yong-Woo Jeon[3], Soonkeum Han[4] and Dong-Cheon Seo[5]

1 Hyena Co., Bucheon, South Korea

2 Hyena Co./The University of Seoul (UOS), Seoul, South Korea

3 Korea Testing Laboratory (KTL), Seoul, South Korea

4 Korea Center for Sustainable Development (KICSD), Seoul, South Korea

5 Jeollanam-do Environmental Industries Promotion Institute (JEIPI), Jeollanam-do, South Korea

*Address all correspondence to: dhlee@uos.ac.kr

References

[1] Akihiko O. History of World Skyscrapers. Japanese, Tokyo: Kodansha; 2015. Translated by Lee Ki-bae into Korean and published by Miseum, Seoul; 2024. (In Korean)

[2] Korea Ministry of Land, Infrastructure and Transport. Statistics of Housing Construction for Construction Consent in 2022. 2023. (In Korean). Available from: https://stat.molit.go.kr/portal/cate/statView.do?hRsId=31&hFormId=626&hSelectId=626&hPoint=00&hAppr=1&hDivEng=&oFileName=&rFileName=&midpath=&sFormId=626&sStart=2022&sEnd=2022&sStyleNum=125&settingRadio=xlsx [Accessed: July 8, 2024]

[3] Tchobanoglous G, Theisen H, Vigil S. Integrated Solid Waste Management: Engineering Principles and Management Issues. 1st ed. Singapore: McGraw Hill; 1993. 10-15 pp

[4] Kaza S, Yao LC, Bhada-Tata P, Van Woerden F. What a Waste 2.0: A Global Snapshot of Solid Waste Management to 2050. Urban Development. Washington, DC: World Bank; 2018. Available from: http://hdl.handle.net/10986/30317

[5] Kenny S, Stephenson J, Stern A, Beecher J, Morelli B, Henderson A, et al. From Field to Bin: The Environmental Impacts of U.S. Food Waste Management Pathways (Part 2). Washington, DC: U.S. Environmental Protection Agency (EPA); 2023

[6] Mayor Adams Announces Roadmap for Nation's Largest Compost Collection Program, Including Achieving Decades-Long Goal of Providing Curbside Service to Every New York City Resident; 2023. Available from: https://www.nyc.gov/office-of-the-mayor/news/084-23/mayor-adams-roadmap-nation-s-largest-compost-collection-program-including-achieving#/0 [Accessed: July 1, 2024]

[7] Kjerstadius H, Haghighatafshar S, Davidsson Å. Potential for nutrient recovery and biogas production from Blackwater, food waste and greywater in urban source control systems. Environmental Technology. 2015;**36**:1707-1720. DOI: 10.1080/09593330.2015.1007089

[8] Korea Ministry of Environment, Korea Environment Corporation. Status of Nationwide Waste Generation and Treatment in 2022; 2023. (In Korean). Available from: https://www.recycling-info.or.kr/rrs/stat/envStatDetail.do?menuNo=M13020201&pageIndex=1&bbsId=BBSMSTR_000000000002&s_nttSj=KEC005&nttId=1416&searchBgnDe=&searchEndDe [Accessed: July 8, 2024]

[9] Korea Ministry of Environment. Types of Waste Treatment Facilities. Annex 3 of the Enforcement Ordinance, Waste Management Act; 2023. (In Korean). Available from: https://www.law.go.kr/%EB%B2%95%EB%A0%B9/%ED%8F%90%EA%B8%B0%EB%AC%BC%EA%B4%80%EB%A6%AC%EB%B2%95%EC%8B%9C%ED%96%89%EB%A0%B9 [Accessed: July 8, 2024]

[10] Seoul. Amount of sewage generated per person per day. In: 63rd Seoul Statistical Yearbook. 2023. (In Korean). Available from: https://stat.eseoul.go.kr/statHtml/statHtml.do?orgId=201&tblId=DT_201004_O120017&conn_path=I2&obj_var_id=&up_itm_id= [Accessed: July 1, 2024]

[11] Vesilind PA, Morgan SM, Heine LG. Introduction to Environmental

Engineering. 2nd ed. Tarun Gupta: Cengage Learning; 2004. pp. 347-349

[12] Korea Rural Development Administration. Notice for Fertilizer Official Specification; 2024. (In Korean). Available from: https://www.law.go.kr/LSW/admRulLsInfoP.do?admRulId=35327&efYd=0#J2741527 [Accessed: July 12, 2024]

[13] Korea Ministry of Agriculture, Food and Rural Affairs. Notice for Feed Standards and Specifications; 2024. (In Korean). Available from: https://www.law.go.kr/admRulSc.do?menuId=5&subMenuId=41&tabMenuId=183&query=%EC%82%AC%EB%A3%8C#liBgcolor10 [Accessed: July 12, 2024]

[14] Jung J, Pyun K, Yoonsuk J. Secret of 20,000 Tons per Day of Food Waste: A Quarter of Food Is Thrown Away before it Is Eaten; 2022. (In Korean). Available from: https://www.joongang.co.kr/article/25041150 [Accessed: July 1, 2024]

[15] Ham GY, Lee DH, Matsuto T, Tojo Y, Park JR. Simultaneous effects of airflow and temperature increase on water removal in bio-drying. Journal of Material Cycles and Waste Management. 2020;**22**:1056-1066. DOI: 10.1007/s10163-020-01000-x

[16] Park JR, Lee DH. Effect of aeration strategy on moisture removal in bio-drying process with auto-controlled aeration system. Drying Technology. 2021;**40**:2006-2020. DOI: 10.1080/07373937.2021.1912080

[17] Tchobanoglous G, Theisen H, Vigil S. Integrated Solid Waste Management: Engineering Principles and Management Issues. 1st ed. Singapore: McGraw Hill; 1993. 7-8 pp

[18] Tchobanoglous G, Theisen H, Vigil S. Integrated Solid Waste Management: Engineering Principles and Management Issues. 1st ed. Singapore: McGraw Hill; 1993. 15-17 pp

[19] Japan Sewage Works Association. Performance Standards for Disposer Wastewater Treatment Systems for Sewerage Systems (Draft); 2013. (In Japanese). Available from: https://www.jswa.jp/wp2/wp-content/uploads/2018/03/e00b34b8444b08b85960371ae40fcc611.pdf [Accessed: July 9, 2024]

[20] Achinas S, Euverink GJW. Theoretical analysis of biogas potential prediction from agricultural waste. Resource-Efficient Technologies. 2016;**2**:143-147. DOI: 10.1016/j.reffit.2016.08.001

[21] Narisetty V, Adlakha N, Singh NK, Dalei SK, Prabhu AA, Nagarajan S, et al. Integrated biorefineries for repurposing of food wastes into value-added products. Bioresource Technology. 2022;**363**:127856. DOI: 10.1016/j.biortech.2022.127856

[22] Seocho-gu. 35th Seocho Statistical Yearbook; 2023. Available from: https://www.seocho.go.kr/site/seocho/07/10705030300002017072804.jsp [Accessed: July 1, 2024]

[23] Korea Ministry of Environment, Korea Environment Cooperation. The 6th (2021-2022) Korea Waste Statistical Survey; 2022. 114 p

[24] Nownsurvey: Survey on Household Waste Emissions - Analysis of Survey Results - How Many Times a Week Do you Go out to Throw Away Food Waste? 2020. Available from: https://www.nownsurvey.com›issue›view›qazvnyhuik [Accessed: June 29, 2024]

[25] Korea Environment Cooperation, Korea Ministry of Environment. How to

Practice Low-Carbon Life. Available from: https://cpoint.or.kr/user/info/wisdom.do?tab=2 [Accessed: June 29, 2024]

[26] User's manual of Hina Clean-up System provided by Hyena Co. Ltd for the certification test in 2017

[27] Yoo KY. A Study on Construction and Management of Food Waste Treatment Facilities. Seoul: Seoul Institute; 2001

[28] Korea Ministry of Environment. Report on Evaluation Results of Waste Treatment Business and Waste Treatment Facility Installation and Operation Status; 2021. 50 p

[29] Helliwell JF, Layard R, Sachs JD, De Neve J-E, Aknin LB, Wang S. World Happiness Report 2024, UN Sustainable Development Solutions Network (SDSN); 2024. Available from: https://happiness-report.s3.amazonaws.com/2024/WHR+24.pdf [Accessed: July 15, 2024]

[30] Landes X, Unger C, Andsbjerg K, Frank K, Wiking M. Sustainable Happiness: Why Waste Prevention May Lead to an Increase in Quality of Life. Danish Ministry of the Environment 2015. Available from: https://static-curis.ku.dk/portal/files/130685640/Sustainable_Happiness.pdf

Section 3

Industrial Waste and Pollution Management

Chapter 5

Shredding of Lithium-Ion Batteries: Overview and Industrial Perspective

Marcelo Oliveira, Bárbara Abreu and Henrique Costa

Abstract

The exponential growth of electrification and digitization is massifying the production and use of lithium-ion batteries, which eventually will reach their end of life. The recycling of these batteries is essential for environmental reasons, but it also shows a strong economic impact. In fact, Li-ion batteries contain critical raw materials in higher concentrations than those found in natural resources. Therefore, lithium-ion battery recycling is an essential and rapidly expanding sector. The recycling process typically involves a shredding step followed by sieving to recover the "black mass"—a powder rich in lithium, cobalt, nickel, and manganese, depending on the battery chemistry. However, shredding an energy storage device containing toxic compounds is challenging, and the most crucial part of the process. Hence, the proposal entails the authoring of a chapter outlining and analyzing various shredding technologies for Li-ion battery recycling, covering the advantages and disadvantages of each from an industrial perspective.

Keywords: recycling, shredding, black mass, lithium batteries, electrolyte

1. Introduction

The exponential growth of electrification and digitization is massifying the production and use of lithium-ion batteries, which eventually will reach their end of life. The recycling of these batteries is essential for environmental reasons, but it also shows a strong economic impact. In fact, Li-ion batteries contain critical raw materials in higher concentrations than those found in natural resources. Therefore, lithium-ion battery recycling is an essential and rapidly expanding sector.

The recycling process typically involves a shredding step followed by sieving to recover the "black mass"—a powder rich in lithium, cobalt, nickel, and manganese, depending on the battery chemistry.

Safety must be considered in all aspects. The electrolyte poses a risk of fire and explosion, while also representing a potential gas inhalation hazard of toxic volatile organic compounds. Additionally, the black mass presents an inhalation risk of particles with carcinogenic potential. The most recent batteries also include carbon nanotubes in their formulation, which present an additional challenge.

Therefore, the technology involved in shredding must be capable of addressing all these issues while simultaneously delivering an effective and safe solution to make this activity socially and economically sustainable.

This chapter explores various industrial primary crushing processes for lithium batteries from both a safety and process performance perspective.

2. Safety issues

Dismantling and crushing devices manufactured to store large amounts of energy is a challenging task. Additionally, the raw materials that make up batteries are potentially hazardous [1, 2]. Therefore, in a chemical recycling facility, potential hazards include:

- Biological risk due to possible inhalation of gases and vapors during crushing or drying, such as electrolyte compounds or degradation products of the conductive salt with water and organic compounds.

- Biological risk due to possible inhalation of black mass dust resulting from the transport, sieving, and packaging of black mass.

- Fire risk during crushing and drying due to the presence of flammable compounds in the electrolyte composition.

- Explosion risks in certain scenarios if chemicals are not properly contained and handled.

- Risks of corrosive chemical products resulting from secondary reactions that can occur during processing.

For greater awareness of the hazards involved, **Table 1** provides safety information for each individual compound.

Considering the intrinsic characteristics of each compound, it is necessary to adhere to the exposure limits, which are tabulated in **Table 2**. It is important to note that there are currently no established limits for the specific compounds of interest,

Compound	CAS	Origin	Hazard statement	Hazard symbol
Lithium cobalt oxide	12,190–79-3	Cathode	H351, H360Fd, H410	
Lithium nickel manganese cobalt oxide	346,417–97-8	Cathode	H317, H351	
Lithium iron phosphate	15,365–14-7	Cathode	—	
Lithium manganese oxide	12,057–17-9	Cathode	H302, H332, H413	
Lithium titanate	12,031–82-2	Cathode	H302, H315, H319	
Lithium nickel cobalt aluminum oxide	193,214–24-3	Cathode	H317, H351	
Carbon black	133–86-4	Anode	H251, H351, H372	
Graphite	7782-42-5	Anode	H319, H335	
Lithium hexafluorophosphate	21,324–40-3	Electrolyte	H302, H314, H318, H335, H372	
Ethyl methyl carbonate	623–53-0	Electrolyte	H225	
Ethylene carbonate	96–49-1	Electrolyte	H302, H319, H373	
Fluoroethylene carbonate	114,435–02-8	Electrolyte	H302, H315, H317, H319	
Copper foil	7440-50-8	Anode	—	
Aluminum foil	7429-90-5	Cathode	H228, H261	
Hydrogen fluoride	7664-39-3	Degradation Product	H300 + H310 + H330, H314	

Table 1.
Chemical compounds found on lithium-ion batteries and respective hazard statement [1, 2].

Compound	Long-term exposure limit (8 h TWA*)		Short-term exposure limit (15 min TWA*)		Ref.
	ppm	mg/m^3	ppm	mg/m^3	
Lithium cobalt oxide (LCO)	—	0,1	—	—	[2]
Lithium nickel manganese cobalt oxide (NMC)	—	0,1	—	—	[2]
Lithium iron phosphate (LFP)	—	5	—	—	[2]
Lithium manganese oxide (LMO)	—	0,5	—	—	[2]
Lithium titanate (LT)	—	0,1	—	—	[2]
Lithium nickel cobalt aluminum oxide (NCA)	—	0,1	—	—	[2]
Carbon black	—	3,5	—	7	[2]
Graphite	—	—	—	—	[2]
Inhalable dust	—	10	—	—	
Respirable dust	—	4	—	—	
Copper dust	—	1	—	2	[2]
Aluminum dust	—	—	—	—	[2]
Inhalable dust	—	10	—	—	
Respirable dust	—	4	—	—	
Lithium hydroxide	—	—	—	1	[2]
Hydrogen fluoride	1,8	1,5	3	2,5	[2]
Lithium hexafluorophosphate	Not established				[2]
Dimethyl carbonate	—	100 mg/L	—	—	[3]

TWA: total weight average.

Table 2.
Work exposure limits of compounds found during the recycling of lithium-ion batteries.

for example, lithium cobalt oxide (LCO). Although a specific limit does not exist for LCO, it is possible to establish a provisional limit until more specific information is released by the responsible authorities, since it is composed of 93% cobalt oxide by mass for which there is an official exposure limit [2].

3. Overview of the process

The shredding process of lithium-ion batteries is a size reduction operation that enables the subsequent separation of components. Its ultimate function is to concentrate the cathode and anode components into a finer fraction and other materials into a coarser fraction, thus allowing for separation by simple sieving based on particle size.

The primary shredding stage aims to grind the battery into smaller particle sizes, thereby destroying all associated battery, module and cell external structures. This is typically done using shaft or rotor shredders [4, 5]. However, since primary shredding is not entirely effective, it is common to employ more stages of crushing to maximize the disaggregation of active materials from their respective copper and aluminum sheets – secondary shredding.

For primary shredding, there are many different types of shredders, however, due to their overwhelming presence, there is a consensus that the best option is to use four-shaft crushers, although other types of crushers are also employed to some extent.

The size and number of shafts, the spacing between them, their geometry, and the material used in their construction are of critical importance for ensuring good performance and proper maintenance of the machine over time.

The purchased equipment must meet the requirements for crushing speed and the maximum allowable size of the battery/module, i.e., whether the operator intends to crush modules or whole batteries.

When considering crushing step capacity, one must consider the speed of the electrolyte drying process. In fact, these two sequential processes have significantly different speeds; crushing is much faster than drying. Therefore, the crusher can only be fed at a rate that the drying process can handle, meaning the limiting step is the drying. A solution to increase plant production is to install two or more dryers so that a discontinuous process like drying can operate semi-continuously. That is, while one dryer is operating, the other is discharging and then being fed with more crushed material. Alternatively, instead of a second dryer, a more economical solution would be buff container to receive crushed wet material, that will be there waiting for the completion of the drying process, although with all the associated limitations.

Patents US11050097B2 and US20210175556A1 state a preference for obtaining 90% of the crushed material with a size below 20 mm, mentioning screens of 20, 30, and 50 mm [6, 7]. These sizes also go in line with the recommendations given by various equipment manufacturers, therefore the optimal size should be within this range.

The optimal screen size will depend mostly on the type and models of batteries that the recycler intends to crush. For instance, a module containing cylindric cells will require a smaller screen, since the diameter of each cylindric cell is typically lower than 20 mm, making a 10 mm screen potentially more effective. Furthermore, the recovery rate of active materials can vary significantly between different module models; one model might achieve high recovery rates, while another might perform poorly, depending on how the batteries are manufactured and the raw materials and additives used. Although R&D in the field has led to significant advancements in battery durability and performance, these improvements have also made the recycling process more challenging.

The secondary shredding stage aims to increase the yield of black mass extraction, i.e., to remove the maximum amount of active material still adhered to the electrode sheets and the separator even after primary shredding. This stage is conducted using mills that apply predominantly abrasive and/or impact forces, such as a hammer mill [4, 5]. The agitation of the crushed material in specialized equipment can also be considered a mechanical mechanism for the disaggregation of cathode and anode powder, even though it does not involve associated direct size reduction [4, 5].

Considering the nature of the shredding operation, equipment suppliers must ensure safety regarding fire and explosion control, and also adhere to the safety limits presented in **Table 2**. All processes with a strong focus on safety perform the shredding step using ATEX-certified equipment and in a closed system so that dust and gases can be diverted away from the atmosphere.

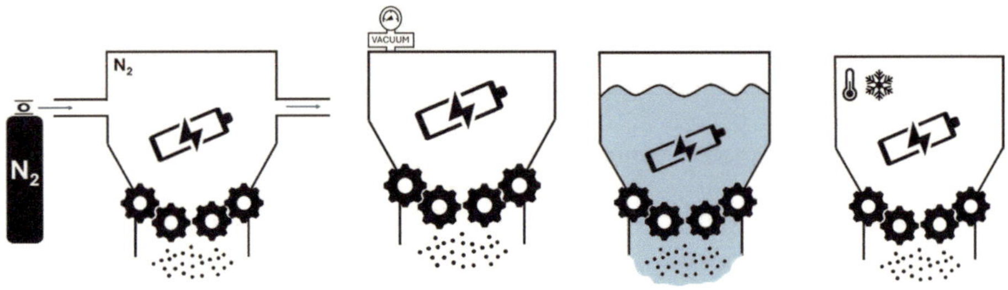

Figure 1.
Different technologies to shred a lithium-ion battery (a) Inert (b) Vacuum (c) Wet and (d) Cryogenic.

The majority of processes control the atmosphere inside the equipment to minimize the possibility of ignition or explosion by reducing the oxygen content, for example, by using an inert nitrogen atmosphere. A variation of this method includes using vacuum or other inert gases. Additionally, some processes use fluids to maximize heat dissipation and minimize contact with oxygen, with the advantage of simultaneously leaching the conductive salt.

Thus, four shredding technologies have been identified from a safety perspective, namely inert shredding, vacuum shredding, wet shredding, and finally cryogenic shredding (**Figure 1**).

4. Shredding technologies

4.1 Inert shredding

This technology involves shredding the samples in a controlled atmosphere using nitrogen. An inert atmosphere enhances the overall safety of the process by minimizing the amount of oxygen, which reduces the probability of fire and explosion. An important factor is the maximum allowable oxygen concentration in the shredding and drying chamber, which some authors state a maximum allowable concentration of 3% [4, 5]. Additionally, a continuous flow of nitrogen will prevent the accumulation of flammable and toxic gases from the evaporation of the electrolyte, particularly relevant when working with solvents with low flash point, such as dimethyl carbonate (DMC) and ethylmethyl carbonate (EMC).

Despite the clear advantages of the technology, inert gas shredding raises several considerations that must be addressed in the equipment design. For example, the batteries must be sent to an intermediate compartment, called airlock, that separates the external air atmosphere (21% oxygen) from the crusher atmosphere (<3% oxygen) by air-tight gate valves [4]. Equipment suppliers must find ways to ensure the system is airtight with optimal sealing throughout its working life span. Additionally, the pressure inside the chamber should be slightly higher than the ambient pressure to prevent external oxygen from migrating into the shredder chamber. To enhance safety, an additional preventive measure is to install an oxygen sensor that will control the operation of the machine, allowing the equipment to work only when the oxygen concentration is below the target setpoint.

Since the maximum allowable oxygen concentration is relatively high compared to the concentration of commercial nitrogen gases, the operator can choose from

Technology	Maximum purity of Nitrogen (%)	Maximum nitrogen consumption (m^3/h)
Cylinder	99,9999	1
Liquid nitrogen	99,9999	10
Membrane	99,5	1000
Pressure swing adsorption	99,9999	10,000
Cryogenic nitrogen plant	99,9999	100,000

Table 3.
Capacity and purity of nitrogen produced by different process [8, 9].

different techniques for nitrogen production, benefiting from local production. There are three technologies for nitrogen purification: membrane separation, Pressure Swing Adsorption (PSA), and conventional distillation [8, 9].

In **Table 3**, it is possible to assess both the purity of produced nitrogen and the production capacity of each technology.

The cryogenic distillation process is only viable for high flow rates, making it clearly unsuitable for a battery recycler. Given that the purity achieved by any technology is acceptable, the option with the lowest investment is membrane technology, offering a solution with lower space requirements and cheaper installation costs [12, 13].

There is a variation of this shredding method that uses carbon dioxide instead of nitrogen. This method was initially explored by Recupyl, a company based in France. Instead of nitrogen, it uses carbon dioxide (CO_2) as the inert gas. The use of CO_2 in recycling processes can lead to a reaction with lithium compounds to form lithium carbonate (Li_2CO_3), which tends to occur at elevated temperatures. The formation of Li_2CO_3 can be desirable in the recycling process when the goal is to recover lithium for subsequent reuse [10]. More detailed information about this process can be found in the company's patent, specifically patent US7820317 [11].

- At this moment, nitrogen shredding is the recommended technology for a potential recycler looking to enter the market. It is a process with extensive know-how, both in literature and at the industrial level, and there is a larger number of equipment suppliers available, particularly from Germany. Although there is not yet a definitive guide for the composition of black mass required to have commercial value, since inert crushing with 4-shaft shredder is the most used process, it is likely that the black mass produced this way will be of comparable quality to what is currently available on the market.

4.2 Vacuum shredding

To the best of our knowledge, vacuum shredding of lithium batteries is a technique that has not yet been implemented on an industrial scale, although there are clear advantages that could in the future raise interest in this technology.

The concept involves minimizing the total amount of gases within the shredding chamber, whether they are combustible, oxidizing, or inert. Patent WO2023165780A1 states that an explosion of this type could increase the pressure by a factor of 10 compared to the system's initial pressure. Thus, if the chamber pressure is ten times lower than the ambient pressure, the pressure overload of the system and the release of gases out of the chamber are prevented. Therefore, operating below 100 mbar would be a safe option, preferably around 5 mbar [12], since at such low pressure the quantity of oxygen and other gases is negligible.

Vacuum crushing has additional advantages. First, it avoids high consumption of nitrogen compared to inert shredding; second, electrolyte drying should be faster, as, unlike the nitrogen method, vacuum shredding already occurs under conditions that promote the evaporation of major electrolyte compounds, such as dimethyl carbonate (DMC), meaning the drying process begins as soon as the crushing starts. This is an advantage since a lower time in the shredding chamber decreases the probability of electrolyte degradation.

Decomposition reactions occur due to the presence of the electrolyte's conductive salt ($LiPF_{x\backslash}$), electrolyte's organic solvents and residual water and are exacerbated by increased temperature. Therefore, whenever the material is subjected to temperatures above 80°C, some extent of decomposition is expected [6, 7, 12, 13]. The inevitable increase of temperature during shredding is one of the possible justifications for higher degradation of the electrolyte residue, therefore is important to separate it as soon as possible.

Additionally, because the evaporation is faster and occurs as soon as the battery is shredded, it should decrease significantly the total process time—as discussed previously, drying is the slowest step of the process. According to manufactures, about 80% of the electrolyte is removed during the shredding stage, which is higher than in inert gas technology.

One important aspect is to prove that this equipment can maintain the vacuum tightness with prolonged continuous operation during many production years, since it will affect vastly the evaporation rate of this process.

- Vacuum crushing is currently the only technology that is not found at industrial scale, although it does not seem to be particularly hard to scale-up. Its main advantages are as follows:
  - Faster removal of electrolyte, as the shredding is done under vacuum.
  - The electrolyte residue is expected to have a cleaner profile. Depending on the degree of purity, this could facilitate the valorization of this residue, a topic that has not yet been fully developed in the industry.
  - Savings on nitrogen, as in this case, only the vacuum pump requires nitrogen for protection.

4.3 Wet shredding

Wet shredding uses liquids to deactivate the cells during the process without relying on an inert gas or vacuum to control the oxygen concentration in the atmosphere. In fact, the liquid itself acts as a barrier between the fuel and the oxidizer and also serves as a means of mitigating sudden temperature increases during the shredding stage [2]

Another associated advantage is the density-based separation of all components that are less dense than the liquid used in the shredding process, such as plastics, electrode separators, and thermal insulation materials. These will float in the liquid, facilitating their separation. However, this material stream contains some black mass, leading to a decrease in process yield, potentially accounting for 3–7% of the black mass according to patent WO2022/246570 A1 [14]. As a result, an additional washing and filtration process is required to recover this material [14].

The major advantage of this method compared to others lies in its potential for process intensification, where shredding occurs simultaneously with possible separation via leaching. Using a liquid, it is possible to remove the conductive salt ($LiPF_6$) and other electrolyte solvents from the mixture, depending on the fluid used. This is clearly an advantage over the other two shredding methods, as it is impossible to remove the conductive salt through distillation, and high-boiling-point organic solvents can be removed only by using very high temperatures, which may lead to greater degradation reactions. However, there is no evidence yet that a wet shredding process can completely remove these compounds.

In general, there are two fluids of interest: water, as it is an economical option with the ability to act as a medium for discharging batteries; and organic solvents, in particular, a low-boiling-point solvent typically found in the electrolyte could be used, such as DMC or EMC, which could increase the efficiency of the leaching process. These solvents would necessarily have an affinity for both the conductive salt and the high-boiling-point solvents present in the electrolyte, such as ethylene carbonate (EC) and propylene carbonate (PC).

In the case of using water, it is advantageous to add salts to facilitate the dissipation of any residual electricity in the battery. For example, salts like sodium hydroxide, calcium hydroxide, or even lithium hydroxide can be used, giving the liquid a pH of 8 or higher. The basic nature of the solution is essential for preventing acidity, as any acid will be immediately neutralized [14].

It is also possible to reduce black mass losses due to dust, as there is less tendency for dust to become airborne given that it is wet [2].

Although this process is beginning to gain some industrial traction, several disadvantages are associated with it, including the necessary treatment of the effluents generated during the process and the potential formation of hydrogen due to the interaction between the aqueous effluent and metal foils, especially aluminum, which is known to react with basic aqueous solutions as follows [2]:

$$2\,Al\ (s) + 6\,H_2O\ (l) \rightarrow 2\,Al(OH)_3(aq) + 3\,H_2(g) \tag{1}$$

The reaction mentioned above is usually slow, as aluminum oxide typically provides protection. However, this protective layer is likely to be removed during the shredding step. Any hydrogen formed is highly flammable, posing a significant safety risk [2].

Additionally, the reaction between metallic lithium and water also leads to hydrogen formation.

$$2\,Li\ (s) + 2\,H_2O\ (l) \rightarrow 2\,LiOH(aq) + H_2(g) \tag{2}$$

Furthermore, the presence of hydroxide salts could trigger another secondary reaction that produces hydrogen, as in the case of a reaction with LiOH or any hydroxide salt used [2].

$$2\,Al\ (s) + 2\,LiOH\ (aq) \rightarrow 2\,LiAlO_2(aq) + 3\,H_2(g) \tag{3}$$

All these reactions are exothermic, potentially increasing the temperature and igniting flammable compounds.

An important note regarding the use of sodium chloride as a salt, which is widely mentioned in the literature for battery discharging, is that it can shorten the lifespan of equipment due to the presence of chlorides, which are particularly challenging to manage in steel and iron-based equipment [15]. Moreover, hydrochloric acid formation could also occur. The metal alloy of the equipment would need to be reconsidered to handle these chlorinated compounds.

From a technical standpoint, using DMC in wet shredding appears to be the best option. Since DMC is used in the composition of electrolytes, it likely has a greater affinity for the conductive salt as well as for other organic solvents, such as EC (ethylene carbonate), which is known to be difficult to remove from black mass without using high temperatures. Therefore, extraction with DMC would result in a purer black mass, free of conductive salt and high-boiling-point solvents.

After shredding, the mixture of the shredded battery and the DMC solution is processed to separate the dissolved material from the remaining components. Various separation methods, such as filtration and subsequent distillation, can be used to recover the DMC to be reused in the process.

However, from an operational standpoint, this process requires additional safety measures, as DMC is an inherently flammable liquid, and its inhalation should be avoided due to its volatile nature. Indeed, no literary evidence was found of a shredding application involving an organic solvent. US20220231350A1 shows a possible solution to extract $LiPF_6$ and high boiling point solvents using organic solvents, although it does not prove that this solution can be used in the shredder step [16].

Wet shredding offers a unique advantage over all other shredding methods, specifically the removal of the conductive salt and high-boiling-point solvents (EC, PC), resulting in a pure black mass.

If water is chosen as the immersion fluid, several safety considerations must be addressed due to the degradation of the conductive salt and the secondary reactions that produce hydrogen. The associated costs of treating the contaminated water must also be considered.

Alternatively, using DMC might make chemical sense, but it is a flammable liquid that would be integrated into the process stage with the highest risk of fire and explosion. Currently, there is no mention of using organic solvents in the primary shredding process, only in later stages for extracting already crushed material.

Due to the safety disadvantages and the ongoing costs associated with effluent generation, this technology is not recommended for shredding. The removal of the conductive salt or high-boiling-point solvents can be accomplished in other ways, specifically only in the final black mass, leading to lower reagent costs since the amount of solids to be treated is smaller.

4.4 Cryogenic shredding

Cryogenic shredding is a methodology that involves lowering the temperature of materials using dry ice, liquid carbon dioxide, or liquid nitrogen, followed by shredding the sample to the desired particle size.

The basis of this shredding method is the solidification of the electrolyte, which prevents the movement of lithium ions and consequently eliminates the electrical current, avoiding battery thermal runaway [17]. Cryogenic freezing is also a viable option for transporting batteries in a critical state, potentially allowing these materials to be transported in containers that are not explosion-proof [18].

Regarding the recommended temperature for freezing batteries, studies show that penetration tests on frozen batteries indicate no significant difference in safety at temperatures of −80°C or lower. Starting from −73°C, the temperature gradually begins to rise, reaching −21°C, where thermal runaway occurs, likely due to the melting points of the electrolyte components, allowing greater freedom of movement for lithium ions [17].

This methodology is employed by Retriev Technologies (formerly Toxco) [19]. According to patent US5888463, the process begins with the dismantling of larger batteries and crushing them in a cryogenic environment using liquid nitrogen, until the batteries reach −195°C [20]. This method is highly effective when a large number of batteries need to be crushed. It also makes the external plastic casing hard and brittle, facilitating the crushing process.

The major disadvantage of this process is the high operational costs associated with using liquid nitrogen to freeze the batteries. Additionally, consideration must be given to what happens to the material after being shredded, as its temperature rises and approaches room temperature. With the successive melting of the various electrolyte solvents, it is expected that the temperature will start to rise, potentially triggering thermal runaway.

Although expired, patent US5888463, formerly held by Retriev Technologies (formerly Toxco), indicates that after the material is shredded, it is subjected to a bath in an aqueous solution of NaOH or LiOH, allowing the process to proceed safely, similar to wet shredding, but with all the respective disadvantages previously discussed [20]. Similarly, Redivivus, which uses a cryogenic process, also includes a step of adding water after the cryogenic crushing process to dissipate any remaining energy in the batteries [21].

Although not officially confirmed, the focus of Toxco's (now Retriev Technologies) 1999 patent may have been on lithium-based batteries with metallic lithium chemistry, where cryogenics might be more advantageous, rather than on current lithium-ion batteries. However, Retriev Technologies holds more recent patents in this field, specifically US8882007 and US8616475, both of which involve the use of water during the crushing process in the form of a spray, indicating a shift toward some type of wet shredding [22, 23].

- Cryogenic crushing is a safe option for shredding batteries, even those in poor condition. However, it is necessary to consider the subsequent stages, as the frozen electrolyte will eventually melt and could cause a short circuit in the shredded material. To mitigate this issue, companies include a step where the shredded material is immersed in water, thereby losing the advantages of this method and introducing the disadvantages associated with wet shredding.

To schematize the information described above, **Table 4** has been drawn up with the main advantages and disadvantages of the shredding technologies in this chapter.

Method	Advantages	Disadvantages
Wet Shredding	• Easier to scale up than dry inert shredding methods • Allows harmful chemicals to be neutralized	• Hydrogen is generated if aluminum is present • Addition of alkali reagent to water increases costs • Alkali ions can be corrosive to process equipment • No electrolyte recovery possible • Additional cost of effluent treatment
Vacuum Shredding	• Cheaper to operate than using inert gas • Can be coupled with a vacuum extraction unit to evaporate and condense electrolyte carbonates • Relatively mature technology used in other recycling operations	• Only suitable for batch processing, making it difficult to scale up • Safety not as high as shredding in an inert atmosphere • Not suitable for shredding charged cells
Inert Shredding	• Mature technology used in recycling of other flammable material • Semi-continuous shredding possible with right feeding mechanism	• More expensive than vacuum shredding in utilities • May need inert conveyance line post shredding to avoid fires/explosions • Not suitable for shredding charged cells and modules
Cryogenic Shredding	• Increases the recovery of valuable components • Low energy consumption in the shredding phase • Low temperatures	• Expensive reagents used • Only suitable for batch processing, making it difficult to scale up • Environmental problems from disposal

Table 4.
Summarize advantages and disadvantages of the shredding technologies described in this chapter [24–26].

5. Technology market overview

Conducting technological surveillance is important to understand the usage trends of each technology and to identify the characteristics of future competitors and their black mass.

To assess the market representation of these technologies, research was conducted on both recycling companies and equipment suppliers that offer recycling lines. Indeed, some technologies are not yet ready for industrial-scale application, while others currently dominate the market, as shown in **Figure 2**.

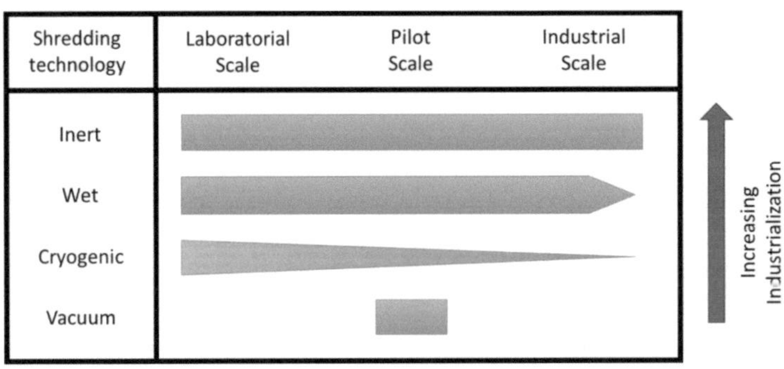

Figure 2.
Representation of each crushing technology according to the processing scale in question.

As previously mentioned, inert gas crushing is the most prominent type of crushing in the industry, with the highest number of recyclers and equipment suppliers at present.

6. Conclusion

The recycling of lithium-ion batteries is at the forefront of industrial technology development. The significant amount of energy stored in these devices presents several safety risks that must be addressed in recycling plants, particularly during the shredding step of mechanical recycling. Among the various risks associated with the process, explosions and fires are the most concerning. Therefore, controlling the atmosphere inside the equipment is a widely applied method to enhance safety.

To date, various technologies are being used industrially, with new ones being developed in response to the continuous evolution of lithium-ion battery composition. In this context, four key technologies have been identified: inert, vacuum, wet, and cryogenic shredding.

Inert shredding, typically using nitrogen, is a safe and mature technology widely adopted by several recyclers and has the highest number of suppliers, with the main limitation being the high nitrogen consumption. Vacuum shredding, while following the same principle as inert shredding, reduces nitrogen consumption and holds promise but has yet to be used industrially. Wet shredding is currently the only technique that effectively removes the conductive salt from the black mass, though the disposal of the resulting residue presents a sustainability challenge. Cryogenic shredding is a safe technology, but it faces several limitations for industrial application. In conclusion, from a safety-oriented perspective, considering the technological expertise and industrial implementation along with the availability of equipment suppliers, nitrogen shredding is the recommended technology for a potential recycler entering the market.

Acknowledgements

This work was developed within the scope of the project "NGS – New Generation Storage" [C644936001-00000045], financed by PRR – Plano de Recuperação e Resiliência under the Next Generation EU from the European Union.

Conflict of interest

The authors declare no conflict of interest.

Author details

Marcelo Oliveira*, Bárbara Abreu and Henrique Costa
CeNTI, Vila Nova Famalicão, Portugal

*Address all correspondence to: moliveira@centi.pt

References

[1] Kaya M. State-of-the-art lithium-ion battery recycling technologies. Circular Economy. 2022;**1**:100015

[2] Sattar A, Cooper L, Dowson M, et al. Considerations in lithium-ion battery recycling. 2023. Available from: https://warwick.ac.uk/fac/sci/wmg/business/transportelec/wmg_-_considerations_in_lithium-ion_battery_recycling.pdf

[3] Li J, Wang G, Xu Z. Generation and detection of metal ions and volatile organic compounds (VOCs) emissions from the pretreatment processes for recycling spent lithium-ion batteries. Waste Management. 2016;**52**:221-227

[4] Diekmann J, Sander S, Sellin G, et al. Crushing of Battery Modules and Cells BT. In: Kwade A, Diekmann J, editors. Recycling of Lithium-Ion Batteries: The LithoRec Way. Cham: Springer International Publishing. 2018. pp. 127-138

[5] Diekmann J, Sander S, Sellin G, et al. Material Separation BT. In: Kwade A, Diekmann J, editors. Recycling of Lithium-Ion Batteries: The LithoRec Way. Cham: Springer International Publishing. 2018. pp. 207-217

[6] Hanisch C, Westphal B, Haselrieder W, et al. Method for the treatment of used batteries, in particular rechargeable batteries, and battery processing installation. US11050097B2. 2021

[7] Hanisch C. Recycling method for treating used batteries, in particular rechargeable batteries, and battery processing installation. US20210175556A1. 2021

[8] Omega Air d.o.o. Ljubljana. Nitrogen and oxygen production. 2023. Available from: https://www.omega-air.si/news/news/nitrogen-and-oxygen-production [Accessed: November 28, 2023]

[9] Van MG. How a nitrogen generator works: PSA vs. Membrane separation technology. 2023. Available from: https://www.ecscorrosion.com/blog/how-a-nitrogen-generator-works [Accessed: November 28, 2020]

[10] Velázquez-Martínez O, Valio J, Santasalo-Aarnio A, et al. A critical review of lithium-ion battery recycling processes from a circular economy perspective. Batteries. 2019;**5**:5-7

[11] Tedjar F, Foudraz J-C. Method for the mixed recycling of lithium-based anode batteries and cells. United States Patent. 2010;**2**:13

[12] Von Loeper B, Hanisch C. Method for recycling components of electrochemical energy stores, and recycling devices therefore. WO2023165780A1. 2023

[13] Han JY, Jung S. Thermal stability and the effect of water on hydrogen fluoride generation in lithium-ion battery electrolytes containing LiPF6. Batteries. 2022;**8**:1-19. DOI: 10.3390/batteries8070061

[14] Biedermen CJ, Ton TG, Lovell MJ, et al. Apparatus for separating materials recovered from batteries. WO2022246570A1. 2022

[15] Shaw-Stewart J, Alvarez-Reguera A, Greszta A, et al. Aqueous solution discharge of cylindrical lithium-ion cells. Sustainable Materials and Technologies. 2019;**22**:e00110

[16] Lipson AL. Method for recovering and recycling electrolyte salts from

lithium batteries. US2022/0231350A1. 2021

[17] Sunderlin N, Colclasure A, Yang C, et al. Effects of cryogenic freezing upon lithium-ion battery safety and component integrity. Journal of Energy Storage. 2023;**63**:107046

[18] Grandjean TRB, Groenewald J, Marco J. The experimental evaluation of lithium ion batteries after flash cryogenic freezing. Journal of Energy Storage. 2019;**21**:202-215

[19] Toro L, Moscardini E, Baldassari L, et al. A systematic review of battery recycling technologies: Advances, challenges, and future prospects. Energies. 2023;**16**: 1-24. DOI: 10.3390/en16186571

[20] McLaughlin W, Adams TS. Li reclamation process. US5888463. 1999

[21] Redivivus. Redivivus. 2022. Available from: https://www.redivivus.tech/

[22] Smith WN, Swoffer S. Recovery of lithium batteries. US8616475B1. 2013

[23] Smith WN, Swoffer S. Process for recovering and regenerating lithium cathode material from lithium-ion batteries. US8882007B1. 2013

[24] Liu J, Bai X, Hao J, Wang H, Zhang T, Tang X, et al. Efficient liberation of electrode materials in spent lithium-ion batteries using a cryogenic ball mill. 2021;**9**(5):106017 (pages 1-8)

[25] Sattar A, Greenwood D, Dowson M, Unadkat P, et al. Automotive lithium-ion battery recycling in the UK. 2020. Available from: https://hvm.catapult.org.uk/wp-content/uploads/2022/06/22350m-WMG-Battery-Recycling-report-v7.pdf

[26] Sivarupan T, Bermingham M, NG C, Sun S, Dargusch M, et al. A review of the use of cryogenic coolnat during machining titanium alloys. 2024;**40**: e00946 (pages 1-25)5

Chapter 6

Intelligent NDT on Used Die/Mould for Remanufacturing

Zezhong Wang

Abstract

Remanufacturing is an important option for industry waste. As a main source of industry solid waste, industry equipment, such as dies and moulds, needs inspections to progress for further remanufacturing processes. To automate inspections such as ultrasonic testing, there are some challenges in robot control, for example, the orientation of transducer and the moving speed. In this study, a model-based method will be used to control the orientation of the end-effector of the robotic arm based on contact force to control the ultrasonic probe to implement robotic UT. A collaborative robotic arm is used to inspect with better safety and precision. A 6 DOF force/torque sensor is used to measure contact force/torque between the end-effector and the surface. A control software platform based on robot operation system (ROS) is established to control the trajectory of robot both in simulation and real world from scratch. Moreover, reinforcement learning (RL) algorithm is implemented to optimise the trajectory, i.e. the orientation control, the moving speed, and position control, during the surface scanning between each waypoint. The proposed method is verified in the control of robot on point-to-point UT, and it can be used in real-world robotic UT.

Keywords: robotic arm, reinforcement learning, trajectory control, ultrasonic testing, industrial equipment

1. Introduction

Manufacturing industry is not only the foundation of human life but also the cradle of innovation. However, manufacturing faces a big problem of environmental pollution. Within the manufacturing sector, there are normally two basic ways to tackle environmental challenges, and the first way is to find new methods to reduce new emissions (e.g. post-process for emissions from factories) or apply new, more environmentally friendly materials in future. The second way is to consume less resource in the current situation. As the manufacturing industry is switching from linear economy (LE) to circular economy (CE) [1], strategies, such as repairing, reconditioning, remanufacturing, and recycling (4R), have been implemented in the industry.

But while more research is on the 4R of the resource of mass products, less researchers focus on the equipment used in the manufacturing industry. There are many types of industrial equipment used to improve efficiency and manufacture more products, for example, important components of the machinery in production line, like dies and moulds. Dies and moulds are both used to cut the raw material or form the shape. Dies can be currently used to produce finer products with new materials, i.e. aluminium and titanium in automotive and aerospace industry [2]. Plastic and composite components can be also manufactured by dies. In the process of die casting, the terminologies of dies and moulds are used interchangeably. The injection moulding was developed rapidly to produce plastic parts [3]. The dies and moulds are used to produce all kinds of products, for example, bottles, shoe soles, and precise high-quality products such as LEGO. Even though additive manufacturing (AM) on metal and non-metal materials developed fast these years, due to the immaturity of the features of AM parts, dies and moulds are currently not replaceable due to their rapid manufacturing speed, high reliability, and relatively low cost.

There are reasons why dies and moulds should be regularly maintenance and remanufactured. At first, the product is durable and has high added value; if the restoration or repairing methods are effective; products have the potential to be leased as a service; the technology can last longer than the life cycle of the products [4]. Cost saving is the biggest benefit of maintenance/remanufacturing on dies/moulds, and it was reported that the remanufactured part only costs 30–50% of the cost of manufacturing new ones [5, 6]. Regular maintenance or remanufacturing can prevent unexpected breakdowns and save downtime. The cost of making new dies or moulds can be saved for the manufacturer. Additional reasons include when the manufacturing of dies/moulds has stopped, or the dies and moulds are high-value and limited manufactured, the maintenance, remanufacturing on the dies and moulds will be necessary. Moreover, the maintenance/remanufacturing of dies/moulds extends the lifespan and improves the quality of manufacturing. It also saves the material used for new parts, which has a positive environmental impact by saving overall material resource usage. It was reported that repairing and reusing worn dies can save material use by up to 50%. The material scrap rate during manufacturing can be also improved [2]. Remanufacturing also provides the potential to modify dies and moulds to accommodate product changes. This flexibility supports the adaptability of tooling to different production scenarios. Considering the reasons above, the maintenance and remanufacturing of dies and moulds is reasonable. According to report, over 80% moulds used in automotive industry undergoes repairing or remanufacturing [7].

The overall remanufacturing processes on dies and moulds are shown in **Figure 1**. During periodic maintenance or break-down maintenance, inspection should be carried out to confirm the state of the components. The evaluation will be made during the inspection to decide whether the die/mould should go back to service. If a defect is detected, the defect should be identified and evaluated to decide if the defect can be repaired from the economic and technical perspective. If the defect can be accepted, the parts can go back to service. If the tool cannot be remanufactured, it will be disposal. If the defect cannot be accepted and needs to be repaired/remanufactured, the core will go to remanufacturing processes. The remanufacturing processes include conventional method, such as welding, and also modern method, such as hybrid manufacturing. The difference between the remanufacturing of dies/moulds and other components is that the surface requirements of dies/moulds are higher than normal parts. Therefore, extra surface finishing processes, such as heat treatment, are necessary before the remanufactured part goes back to service. Re-assembly and testing before service are needed as well. After the remanufactured dies/moulds go back to service, the circular economy of the tools closes its loop.

The runout of the material during manufacturing processes can be large, and the size wear can be over 1.5 mm in experiment [8]. Therefore, the maintenance and

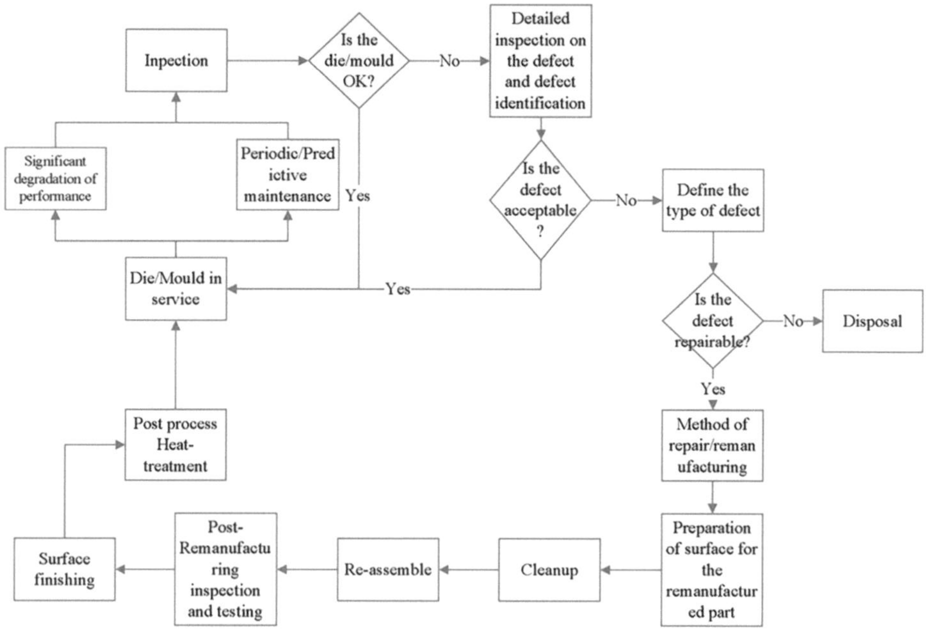

Figure 1.
Flowchart of remanufacturing processes on dies and moulds.

remanufacturing of the dies and moulds is necessary. However, the current maintenance of dies and moulds only includes visual inspection, cleaning, and size evaluation, which will easily leave the subsurface cracks neglected. Moreover, disassembly must be carried out for the complex dies and moulds to make off-line inspection, which will lead to a longer downtime. If an online inspection method can be realised, it will significantly improve inspection efficiency. The purpose to inspect surface and subsurface defects of the manufacturing tools is to evaluate the current condition, even the potential defects in subsurface, and then it can be planned for the future remanufacturing or repairing processes, for example, if we can use hybrid manufacturing method to remanufacture the key components, or the method of repairing/remanufacturing, the variables used in remanufacturing, such as elastic properties of filaments and wires used in AM repairing [9].

Ultrasonic testing (UT) is an important method of non-destructive testing (NDT) method, and it is capable of detecting internal defects, including subsurface cracks. It uses ultrasonic waves to transmit in the material and reflect to the transducer when blocked by some defects. With the development of UT, there have been many ways to implement UT, for example, immersion test, phased array, and contact test. In this study, since the dies and moulds are expensive, they are important to the efficiency of manufacturing. Immersion test is not selected. Since the shapes of dies and moulds are complex, the size of phased array is too large for this type of inspection. Therefore, traditional ultrasonic transducer in contact test is used in this study. The trajectory of the conventional transducer can be planned to realise full cover the surface of object. With the amplitude acquired from the conventional UT, the location, shape, and size of the defects can be reconstructed. With advantages such as reasonable price and flexibility to realise free scanning, the conventional single-element UT probe is used in this study.

In normal industrial UT, most UTs are implemented by human operators. For human operators, the results of inspection heavily depend on the experience and expertise of the operator; in complex cases, even a group of experts is needed to confirm the inspection results and make the decisions for future operations. The stability and efficiency cannot be guaranteed due to many reasons, e.g. physical condition, mental health of the human operator, or environment factors, if the scanning task is on a complex surface or a big area. Moreover, since coupling (normally glycerin) should be applied in contact UT to reduce the attenuation of ultrasonic waves, it can be difficult to move the probe according to the planned path, especially when the scanning surface is complex, i.e. the sharp edges. The academic research on the defects in dies and moulds relies on the off-line inspection methods, which also cost extra downtime. If the subsurface crack is missed and it develops into a surface fatal, a big financial loss, even a fatal accident will be caused by the degraded quality of the products due to defective dies and moulds. Another challenge is that some scanning tasks can be planned according to the original computer-aided design (CAD) model. But for the dies and moulds to be remanufactured, the shape of the parts has been changed from the original designed shape due to harsh working conditions, and therefore, the original CAD model is not always available for planning the UT. As reported, the size of dies and moulds changes along the operation; for example, the size of a die can vary over 1 mm, and the volume change can be over 100 mm^3 after 3000 forgings [10]. The uniform wear value can be over 2 mm [11], which is a big gap for surface scanning. The most importantly, the probe should contact and be normal to the contact surface of the object to guarantee the best results. Otherwise, the ultrasonic wave will be attenuated due to the incidence angle [12]. The volume change after thousands or millions of times operations can be influential to the UT scanning. Therefore, implementing robotic arm with an off-line pre-defined trajectory is hardly satisfactory for this task. On-line trajectory planning optimisation is beneficial. Moreover, as the object is metallic components, the computer vision method will be influenced by the surface reflection of light, and therefore, other methods should be considered.

In this study, a control method is proposed utilising a robotic arm solely with force/torque sensor for automated ultrasonic testing (UT). The pre-defined path is assigned to the robot, but the orientation, z-axis position, and moving velocity of the transducer are controlled by the controller. The implementation will be carried out in simulation at first, and then, it will be transferred to the real world and different shape of objects. In the simulation environment, different types of surface scanning will be trained. An improved compliance controller is implemented in ROS to simulate and experiment on the robot. Reinforcement learning algorithm will be used to help the implementation to adapt to different objects.

2. Simulation model and control strategy

To establish the methodology of this study, a simulation model is built first. The simulation model will help to improve the efficiency of testing and validate the controllers and especially used in the training of reinforcement learning.

In this study, ROS is chosen to be the middleware of the robot simulation, since it contains high-fidelity simulation which is close to the real world. High-fidelity simulation for such surface trajectory for a robotic arm is also a research gap that has been rarely studied by prior research. A Linux laptop with 8Gb RAM is used to run all the codes and connect to the real arm to implement the experiments. UR5e is the robotic arm used to implement the inspection, since the real Universal Robot produces the most

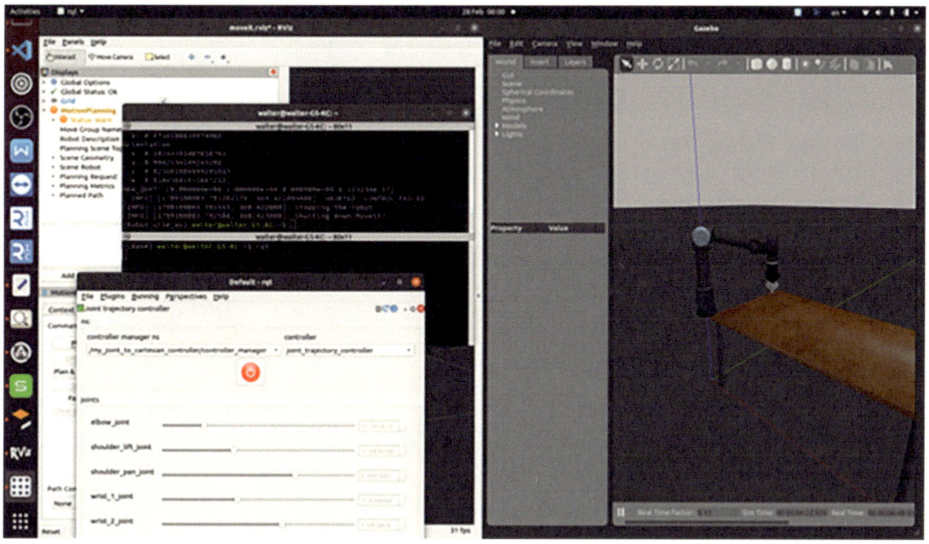

Figure 2.
Screenshot of working scenario in Gazebo.

precise cobot, the repeatability can reach 0.03 mm. Gazebo is the simulation software, and Moveit is the software help planning the trajectory (as shown in **Figure 2**). Except for the robotic arm itself, a 2F-85 Robotiq gripper and a customised transducer holder are modelled and integrated to the simulation model. The gripper is used to grip the transducer to finish the inspection; the transducer holder is to fix the ultrasonic transducer on the gripper, since the transducer is small and difficult to hold by the flat-fingered gripper. With the holder, the transducer goes with the gripper, and the orientation is always the same as the end-effector.

Except for all the hardware settings in the simulation, the most important for inspection is the controller of the robot. The controller is used to maintain the orientation of the end-effector while moving according to the trajectory planning. A compliance controller is used in this study, to apply constant contact force to the surface and also make the end-effector of the robot follow the trajectory planning of the controller. Based on the compliance controller [13], a virtual model is embedded in the code (shown in **Figure 3**), several settings of parameters have been tested, and the final parameters were confirmed as shown in **Table 1**.

With the parameters in the previous table, the proposed controller has been applied to the robotic arm simulation model in the task of scanning on the flat table

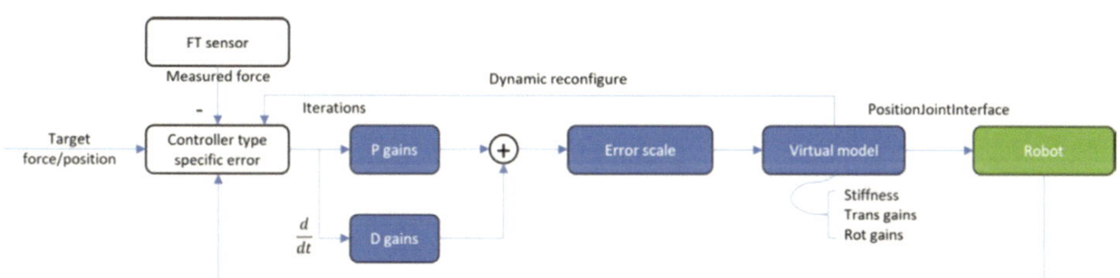

Figure 3.
Details of the solvers in the compliance controller.

	Stiffness		PD		Error scale	Iteration	Result	
	Trans	Rot	Trans	Rot			Average	Derivative
1	500	20	0.005	0.1	0.001	2	36.539	26.019
2	500	20	0.05	0.1	0.001	2	51.558	29.310
3	500	20	0.005	0.1	0.01	2	54.557	3.607
4	500	50	0.005	0.1	0.001	2	57.462	6.063
5	2000	20	0.005	0.1	0.01	2	64.691	7.932
6	1000	20	0.005	0.1	0.001	2	57.023	7.118
7	1000	20	0.005	0.1	0.001	10	60.565	7.464
8	1000	20	0.005	0.1	0.001	20	53.839	7.464
9	1000	20	0.005	0.1	0.001	25	51.062	2.507

Table 1.
Comparison of the control parameters of compliance controller.

surface in the simulation (shown in **Figure 2**). The measured force value in the simulation environment is shown in **Figure 4**.

Except for the compliance controller controlling the contact force, there is another controller responsible for the real-time trajectory. The overall controller of the robot should be used in normal conditions, which is not limited to the contact scenarios. The controller should be able to control UR5e in simulation and real-world, so the RL training results can be transferred between them. The compliance controller is used as a low-level controller as a classic position controller, while the RL controller will be used to train the robot to implement scanning inspection actions. Subsequently, the scanning movements will be transferred to the real world, with the trained RL model and the real-world compliance controller (see **Figure 5**).

In the simulation environment, the scanning on a curved surface is trained by a reinforcement learning model. A package openai-ros [14] is used to connect the middleware-ROS and the reinforcement learning codes in Stable Baselines3 library. After comparison, A2C algorithm was chosen in this study. The simulation scenario is shown in **Figure 6**. In the RL model, current position, orientation, and 6DOF force will be the state of the agent. The next step velocity and orientation adjustment are assigned to the agent as the action. In the RL model, the reward is defined as: if the end-effector

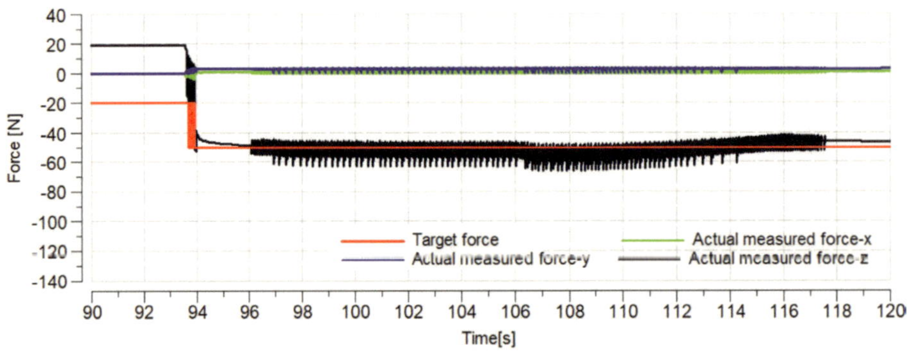

Figure 4.
Force curve of the end-effector during movement.

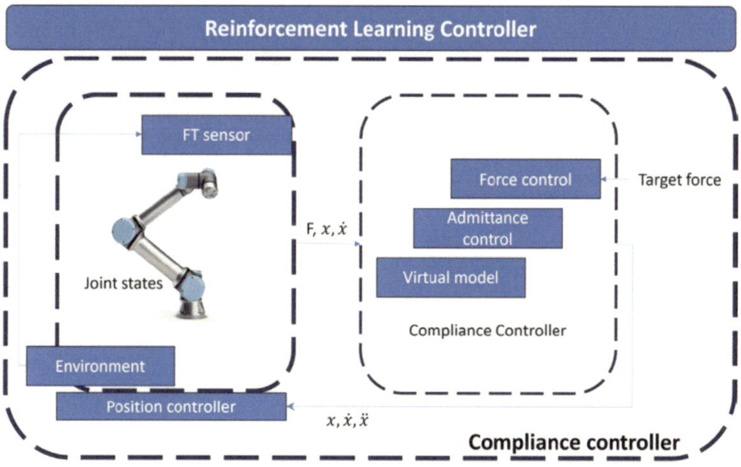

Figure 5.
Control strategy of compliance controller.

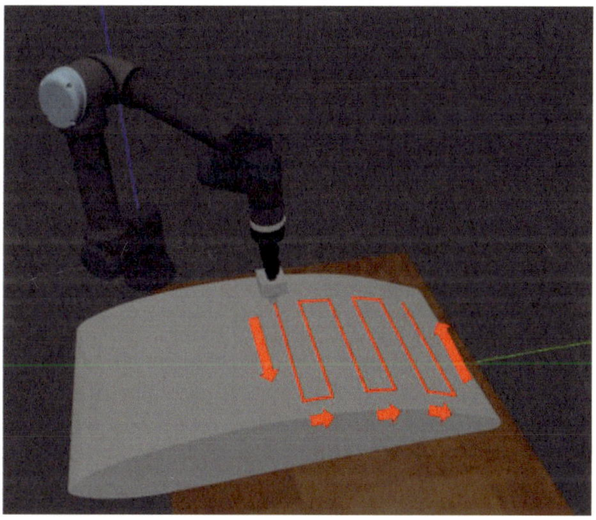

Figure 6.
Raster path planning on wing surface in simulation.

is getting closer to the target, reward +10; if the orientation is close to the normal orientation, reward +10; and if the contact force is close to the target force, reward +10.

The hyperparameters of RL algorithm are tested and compared in the same training scenario. The optimal parameters are listed as follows: Adam optimiser is used as the optimiser in the deep neural network. The learning rate was chosen as 0.001. The discount factor is 0.95. The value function coefficient is 0.8. The reward during the training stage can be seen in **Figure 7**.

The RL model trained the robot scanning on the curved surface, then the trained model is loaded to apply the learned knowledge on a die, which is the target object in this study (shown in **Figure 8**). For safety considerations, the experiment is carried out in simulation environment first. As shown in **Figure 8**, the trained model can maintain good orientation as it is scanning the die. The colour patches in the figure

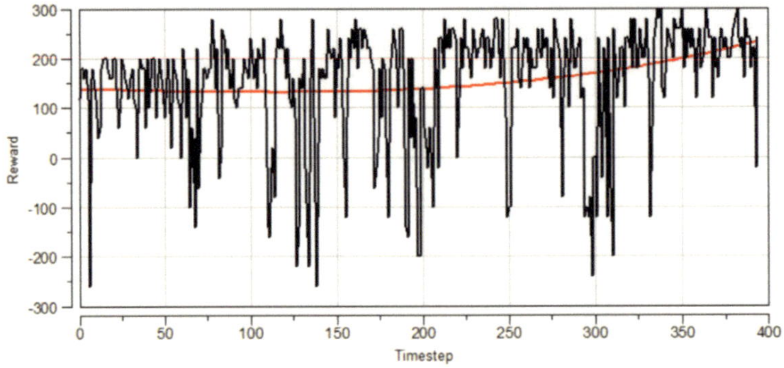

Figure 7.
Reward curve of RL training on curved surface.

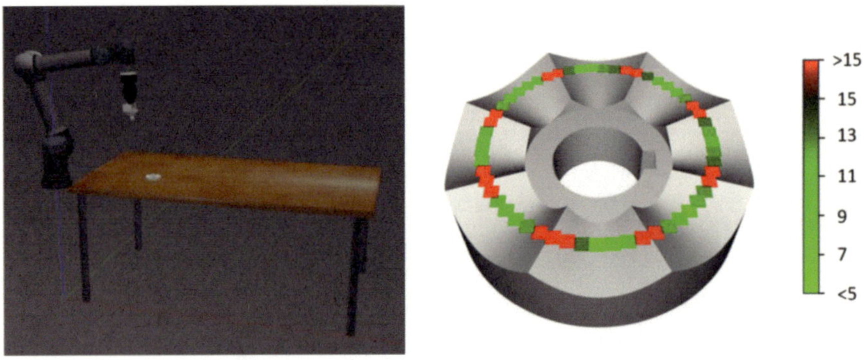

Figure 8.
Accuracy of scanning on die surface.

show the orientation difference between the actual orientation of the end-effector and the normal orientation of the surface. According to the literature, the orientation difference over 15° can affect the scanning of UT. As shown in the figure, the biggest orientation differences happen on the edges of the die, which is more complex situation; moreover, the performance of conventional UT inspection will be not as good as the normal curved surface at the edge. The overall orientation performance is acceptable, i.e. the average of the angle difference is 10.76°. In the experimental stage, the trajectory of simulated robot will be transferred to the real world.

3. Experiments

To validate the proposed method of robotic UT, an experiment is needed. Besides robotic arm and control computer, an ultrasonic probe and data collection device is needed. As a single-element ultrasonic probe is used in this study, the Olympus EPOCH 650 is used as the data collection device. Since the surface of the object is curved, a delay-line probe, V203-RM, is chosen as the probe. All the hardware needed in this thesis is listed in **Figure 9**.

The real object in this study is chosen as a real metallic die (as shown in **Figure 10**). The real object has a small area, but a complex upper surface. For conventional scanning method, the die should be disassembled and access from the other side. In our

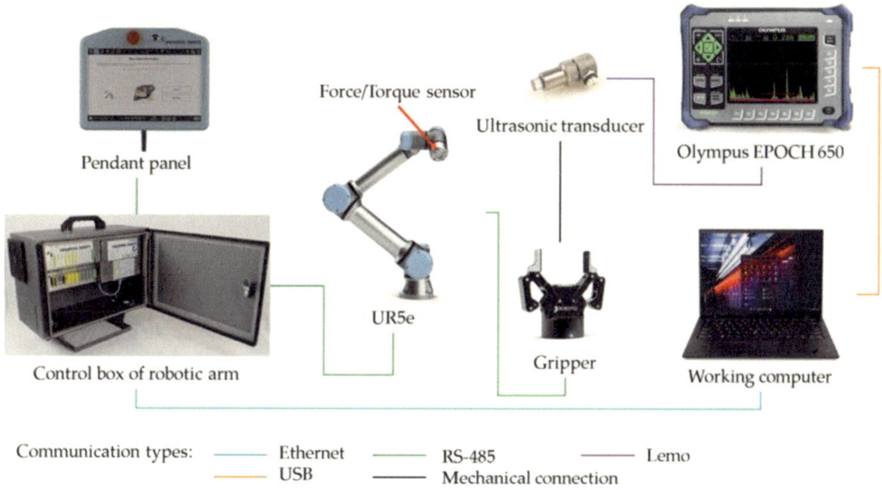

Figure 9.
Hardware used in simulation and experiment in this study.

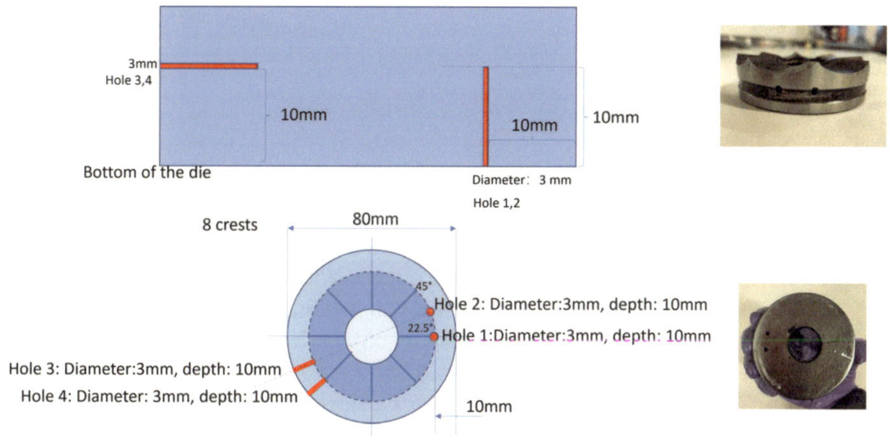

Figure 10.
Test object for the experiment.

case, we do not need to turn the die around if it is assembled on the production line, and direct scanning on the upper surface can be carried out. It can be used to demonstrate the application of the proposed method. Four internal artificial defects, i.e. flat bottom drilled hole (FBH) is carried out on the object to imitate subsurface cracks [15]. The internal structure is as follows: 4 FBHs, i.e. 2 holes are horizontal group, the other 2 are vertical group. Each hole has the same features, 3 mm in diameter, 10 mm in depth. For every 2 holes in a group, one hole is underneath the lowest point of the curved surface, and the other one is underneath the peak point of the curved surface. The design is to verify the capability of robotic UT. For UT of internal defects, the horizontal defects in this case should be simpler to detect, but the vertical defects are difficult. And the defects under the lower point of curved surface should be easier to detect since the upper surface is close to flat. But the defects underneath the peak point of curved surface are difficult to detect, since the upper surface is more complex, and the trajectory of the probe will be more challenging.

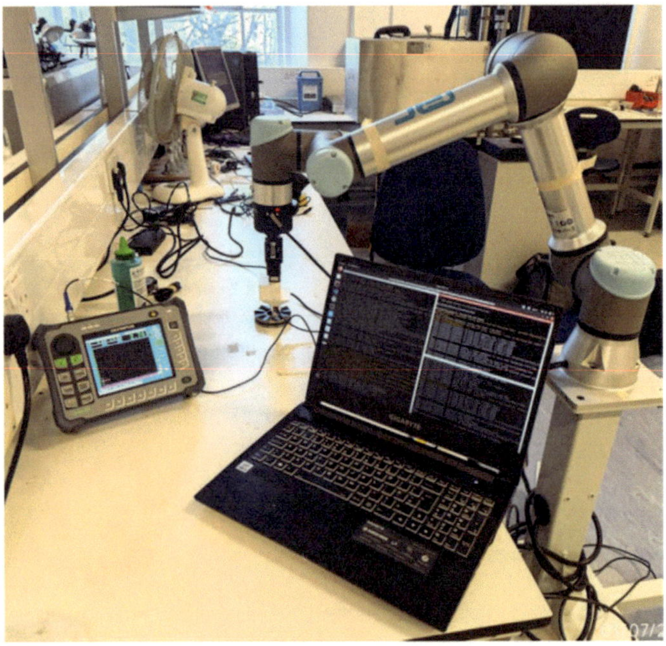

Figure 11.
Photo of experiment scenario with Epoch, PC, robotic arm, and transducer.

The V203-RM probe is a small-sized single-element delay-line probe, which is suitable for curved surface and detection of subsurface cracks. The centre frequency of V203-RM is 10 MHz. The diameter of the element is 0.125 inches (3.175 mm), and the diameter of the case is 0.39 inches (9.906 mm), which is the smallest delay-line probe from Olympus. As shown in **Figure 11**, the probe, the Lemo interfaced wire, and couplant were prepared. Olympus EPOCH 650 is the latest data collection device for single-element probes. The EPOCH 650 is suitable for most kinds of single-element UT inspections, including mechanical testing. The LEMO interface is included, and the A-scan amplitude can be recorded via a USB cable to the control computer. The recorded data is in .csv format. Before planning the experiment, the recording function of EPOCH was tested. The probe was also tested on the real object, and as shown in **Figure 11**, the probe can detect the subsurface defects at 8 mm depth.

To log the data in EPOCH650, a series of software named "GageView" will be used. In Epoch650, live videos can be exported to external storage. After the export, the video is in the format of ".vid", which is in a special code that the video players cannot play. A transformation software "GageView Video" should be used to convert the .vid to .avi format. The video can be presented in the presentation.

4. Results

To verify the working process or robotic UT, the UT on a real die was implemented. It uses compliance controller with target_wrench -50 N, and the path of the end-effector is a circle with a radius of 5 cm. At each position, the thickness and defects were measured with EPOCH650, and the results are shown in **Figure 12**.

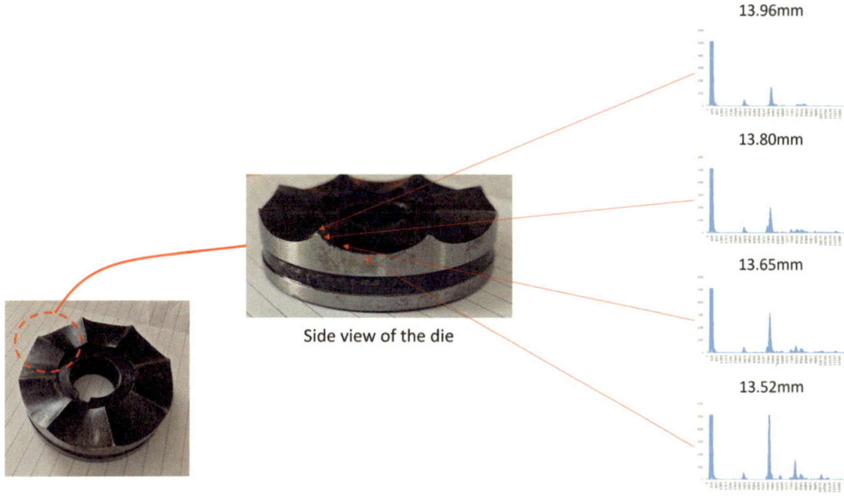

Figure 12.
UT raw data of thickness measurement around the surface of the die.

During the experiment, 20 times scanning was repeated, and the detection of the artificial defects was 16 times. The probability of detection (POD) reached 80%. The overall scanning time is improved significantly.

As shown in **Figure 13**, the amplitude of ultrasonic wave reflection can be read in EPOCH and exported to .csv document, and the curves in the figures are from the data. The two curves in **Figure 13** are only, for example, to compare the result at the position without a defect and the position with a defect. When there is no internal defect, the ultrasonic wave will penetrate through the object and there will be a reflection at the back surface. Since the energy of ultrasonic wave will attenuate, the largest reflection is at the back surface of the object, and the amplitude afterwards will be smaller. As the defects interrupt the ultrasonic wave, there is a small reflection on the EPOCH screen. Single-element conventional ultrasonic probe is used in this study, due to its reasonable price and straightforward usage. To achieve better resolution inspection images in future, a multi-element phased array ultrasonic testing (PAUT) probe can be used; however, the focus of this study is the control of robotic arm, and PAUT was not utilised [16].

The proposed method was compared to manual UT which was implemented by a medium experienced operator. To do the same scanning on the object, the manual

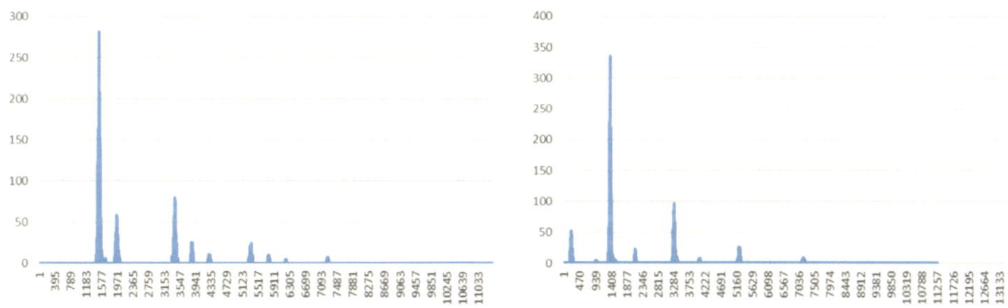

Figure 13.
Ultrasonic wave amplitude curve from the results from EPOCH equipment. Left: data from the position without a subsurface defect; right: position with a subsurface defect.

UT took 4 hours while the robotic UT took half of the time, 2 hours. For accuracy, since the manual UT is without any orientation and position measurement, it cannot be evaluated. The POD of proposed method can reach 80%, while manual method can only reach 50%, and many re-scannings were carried out during the processes. Therefore, it was proved that the proposed method can be more effective and efficient than manual UT.

5. Conclusion

In this study, a control platform for UR5e robotic arm was established. It includes a simulation environment and the implementation in the real world. Automated ultrasonic testing surface scanning task was simulated in the platform and then carried out in the real world. Single-element UT transducer was fixed on the end-effector of robotic arm with a 3D-printed holder. RL algorithm was used to improve the control of robotic arm to make the UT task more reliable and efficient. With the RL algorithm, the trajectory of UT scanning can be improved to achieve better performance. The RL trains the robot model for the assigned tasks, and the trained model can be carried out in different shape in the real world. The RL model was trained in Gazebo first and then transferred to the real world. The implementation in the real world proves that the robotic automatic UT has better efficiency and better adaptability than human manual UT and conventional control method. The method does not need to plan the trajectory off-line on CAD model of the object and can be implemented on any object. The platform established in this study can be used in inspection in manufacturing industry, and it can help inspection in remanufacturing and help improve economic benefits for the entrepreneurs. The traditional entrepreneurs may not be interested in this topic since it is a bit far away from current realistic manufacturing; however, it can be also used in the future research as a basement to develop future robotic contact tasks, such as robotic 3D printing remanufacturing and polishing. With this automated inspection method, the maintenance and remanufacturing of key components in manufacturing industry can be improved efficiently and effectively, and the waste in manufacturing can be reduced further.

Author details

Zezhong Wang
College of Engineering and Physical Sciences, Aston University, Birmingham, United Kingdom

*Address all correspondence to: walterwzz@hotmail.com

References

[1] Ramakrishna S et al. Emerging industrial revolution: Symbiosis of industry 4.0 and circular economy: The role of universities. Science, Technology and Society. 2020;**25**(3):505-525

[2] Bennett J et al. Repairing automotive dies with directed energy deposition: Industrial application and life cycle analysis. Journal of Manufacturing Science and Engineering. 2019;**141**(2):021019

[3] Veligorskyi O et al. Artificial neural network motor control for full-electric injection moulding machine. In: 2019 IEEE International Conference on Industrial Technology (ICIT). Melbourne, VIC, Australia: IEEE; 2019. p. 60-65

[4] Yeo N, Pepin H, Yang S. Revolutionizing technology adoption for the remanufacturing industry. Procedia CIRP. 2017;**61**:17-21

[5] Saboori A et al. Application of directed energy deposition-based additive manufacturing in repair. Applied Sciences. 2019;**9**(16):3316

[6] Thompson S. Handbook of Mould, Tool and Die Repair Welding. Cambridge, England: Elsevier; 1999

[7] Chen C et al. A review on remanufacture of dies and moulds. Journal of Cleaner Production. 2014;**64**:13-23

[8] Lin JC et al. The optimal clearance design of micro-punching die. Journal of Achievements in Materials and Manufacturing Engineering. 2008;**29**(1):79-82

[9] Honarvar F, Varvani-Farahani A. A review of ultrasonic testing applications in additive manufacturing: Defect evaluation, material characterization, and process control. Ultrasonics. 2020;**108**:106227

[10] Hawryluk M et al. Analysis of a forging die wear by 3D reverse scanning combined with SEM and hardness tests. Wear. 2021;**476**:203749

[11] Xu W, Li W, Wang Y. Experimental and theoretical analysis of wear mechanism in hot-forging die and optimal design of die geometry. Wear. 2014;**318**(1-2):78-88

[12] Zhang H, Xu C, Xiao D. Offline correction of tool path deviations for robot-assisted ultrasonic nondestructive testing. Proceedings of the Institution of Mechanical Engineers, Part C: Journal of Mechanical Engineering Science. 2019;**233**(8):2879-2893

[13] Scherzinger S, Roennau A, Dillmann R. Forward Dynamics Compliance Control (FDCC): A new approach to cartesian compliance for robotic manipulators. In: 2017 IEEE/RSJ International Conference on Intelligent Robots and Systems (IROS). Vancouver, Canada: IEEE; 2017

[14] Yadavari H, Tavakol Aghaei V, İkizoğlu S. Deep reinforcement learning-based control of Stewart platform with parametric simulation in ros and gazebo. Journal of Mechanisms and Robotics. 2023;**15**(3):035001

[15] Pau M, Leban B, Baldi A. Simultaneous subsurface defect detection and contact parameter assessment in a wheel–rail system. Wear. 2008;**265**(11-12):1837-1847

[16] Wang Z, Ma X, Xu Y. Weighted multi-element synthetic aperture focusing technique algorithm of ultrasonic non-destructive testing on machinery. In: 2021 26th International Conference on Automation and Computing (ICAC). Portsmouth, United Kingdom: IEEE; 2021

Section 4

Municipal and Community Waste Management

Chapter 7

Overview on Municipal Solid Waste

Mohana Priya Rajendren and Muthuminal Ramu

Abstract

Waste generation around the globe has been enormous in past decades, and also growing rapidly. From statistical data, it is found that every waste generated around the globe is two billion metric tons and it will increase by 70% in 2050. As we all know, municipal solid waste is generated from households, hospitals, and business areas. The items can be paper, cardboard, clothes, kitchen waste, other miscellaneous waste, hospital waste, and a few business scrapes. It is said that each EU citizen produces on average more than 500 kg of MSW a year. Disposing of this effectively involves tedious processing as the material properties differ from each other. The disposing methods are evolving day by day with the help of government and non-government organization. This book chapter proposes to present collective data on municipal solid waste, types of municipal solid waste, policies promulgate with municipal solid waste, effective disposing methods of MSW and how government has established seven-step plans.

Keywords: waste generation, globe, two million metric tons, waste management, disposing methods, municipal solid waste, types of MSW, policies with MSW

1. Introduction

India's metropolitan areas are increasingly challenged to meet the escalating infrastructure demands of a populace becoming increasingly urbanized. Of the 1.21 billion people living in India, 31% are city dwellers, according to the 2011 census. Half of India's population is expected to reside in cities by 2050.

Megacities are a relatively new phenomenon linked to economic, cultural, and technological globalization [1]. India's megacities include Bangalore (6.3 million), Hyderabad (7.7 million), Ahmedabad (6.3 million), (8.4 million), Greater Mumbai (18.4 million), Delhi (16.3 million), Chennai (8.6 million), Kolkata (14.1 million), and Greater Mumbai (18.4 million) [2]. These cities produce large amounts of waste when compared to other states.

As the population grows, the nation's municipal solid waste management (MSWM) has become more difficult due to the enormous amounts of MSW that are produced daily in addition to environmental and esthetic problems. The Central Pollution Control Board (CPCB) reports that during 2014–2015, India produced 1,43,449 tons per day (TPD) of MSW, or 0.11 kg/capita/day on average. About 1,17,644 TPD (80%) of the entire MSW was collected, however, only 32,871 TPD

(22%) were processed or treated. The lack of proper waste segregation at the source, collection, transportation, treatment, and scientific disposal of garbage resulted in environmental degradation and a low standard of living.

The 1990s saw the first signs that MSWM was becoming a serious issue when widespread complaints about inappropriate MSWM practices led to several public interest lawsuits (PILs), which forced the Supreme Court of India to form a committee to investigate the state of MSWM in the nation. The Supreme Court received the report on solid waste management in India's Class I cities from the aforementioned committee in 1999.

The Supreme Court then ordered the Ministry of Environment, Forests, and Climate Change (MoEFCC) to publish the MSW (Management & Handling [M&H]) Rules, 2000, which mandate that all urban local bodies (ULBs) in the nation, including metro cities and class I cities, establish an appropriate waste management system and set a deadline for installing waste processing and disposal facilities by the end of 2003. To offer technical assistance to all ULBs, the Ministry of Urban Development (MoUD) also created an MSWM handbook, which was released concurrently with the MSW (M&H) Rules, 2000.

Funds were allotted under the twelfth and thirteenth Finance Commission Grants in 2005 to enhance MSWM through initiatives such as the Urban Infrastructure Development Scheme for Small & Medium Towns (UIDSSMT) and the Jawaharlal Nehru National Urban Renewal Mission (JnNURM). The National Urban Sanitation Policy, which addressed urban sanitation and included solid waste management as a crucial component, was introduced in 2008.

However, despite positive pilots and accomplishments, the majority of ULBs still struggle with the sustainable financial management of MSWM in addition to adequate and sophisticated collection and transportation systems, technology selection, and disposal techniques. The non-compliance issue is still true after 16 years of the notification of the MSW (M&H) Rules, 2000.

The Swachh Bharat Mission was initiated in 2014 by the Ministry of Urban Development to promote the management of municipal solid waste in urban areas. The Swachh Bharat Mission (SBM) is an initiative of the Indian government that was inspired by the honorable president of India's speech to the Joint Session of Parliament on June 9, 2014, which stated that public spaces filled with trash and homes lacking bathrooms should not be accepted as acceptable living conditions. SBM aims to advance cities as catalysts for economic expansion by enhancing the standard of urban infrastructure, ensuring service levels, and fostering effective governance. SBM seeks to assist cities in creating adequate and modern methods while addressing the difficulties associated with managing solid waste in municipal settings.

The MSW (M&H) Rules, 2000 have also been updated by the Ministry of Environment, Forests, and Climate Change, and are now known as the "Solid Waste Management (SWM) Rules, 2016." The MSWM Manual, 2000 has also undergone parallel revisions by the Ministry of Urban Development. The updated handbook is based on 16 years of knowledge acquired in India following the release of the initial edition in 2000.

India faces a challenge with garbage management. With 1.43 billion people as of mid-2023, India has surpassed China as the country with the highest population in the world. This, along with increased urbanization, has resulted in a steady increase in the country's daily municipal trash production, which surpassed 160 thousand metric tons in the fiscal year 2021. To ensure the proper treatment and disposal of garbage, the nation needs to improve its management systems, as millions of tons of waste still wind up in landfills each year and frequently seep into the environment [3].

Per capita generation of solid waste in different States/UTs is illustrated in **Figure 1**. It has been noted that Delhi generates the most solid garbage per person, followed by Lakshadweep and Mizoram, in that order [4].

Figure 2 shows the percentage of solid waste processed in various States/UTs. The highest proportion of treated solid waste (100%) is found in Chhattisgarh, followed by DDDNH (88.76%), Goa (87.04%), the Andaman and Nicobar Islands (84.7%), and Madhya Pradesh (80.67%), in that order [4].

1.1 What is municipal solid waste?

Commercial and residential garbage produced in municipal or notified areas, whether solid or semi-solid, is called municipal solid trash. Industrial hazardous waste is not included in this category, however treated biomedical waste is.

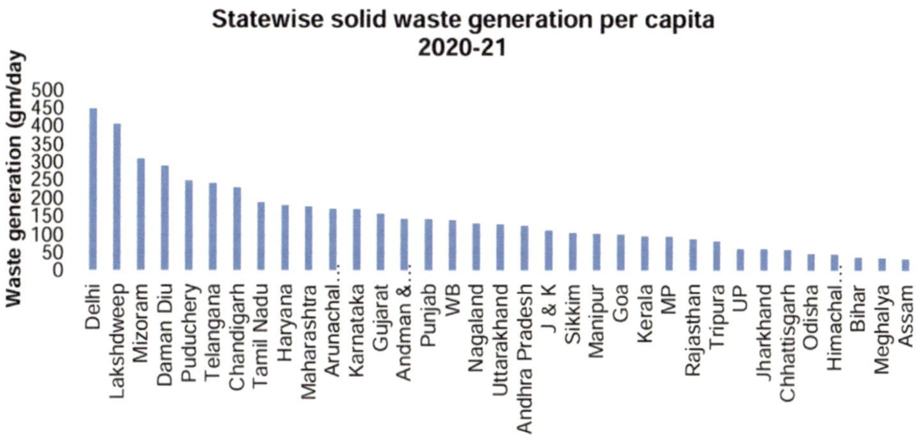

Figure 1.
State wise waste generation data (Source: MSW_AnnualReport_2020-21.pdf (cpcb.nic.in)).

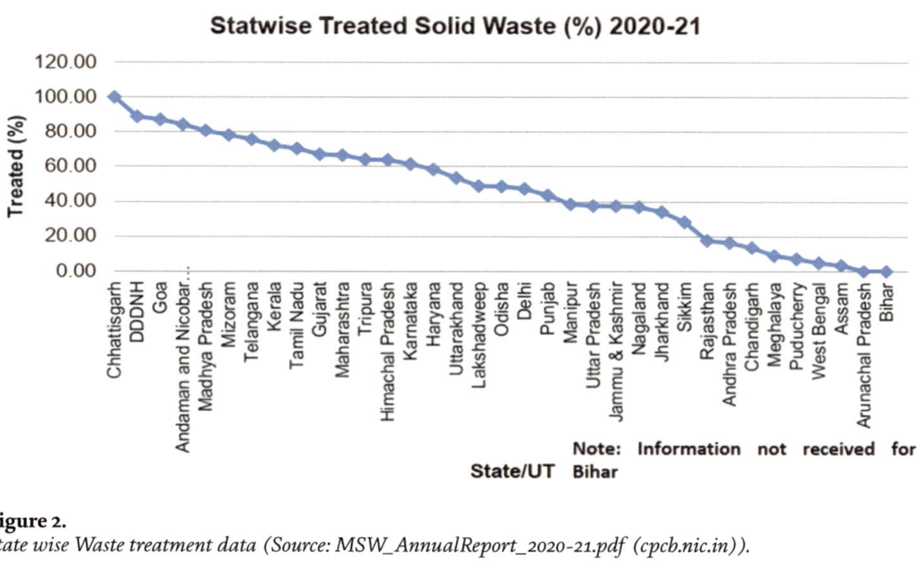

Figure 2.
State wise Waste treatment data (Source: MSW_AnnualReport_2020-21.pdf (cpcb.nic.in)).

2. How municipal solid wastes are collected

It is forbidden to litter municipal solid trash in cities, towns, and urban regions where the state governments have issued a notification. The municipal authorities will implement the following measures to ban littering and encourage compliance:

- Arranging for the door-to-door collection of municipal solid garbage using any of the following methods: using the vehicle's musical bell for scheduling, collecting at regular pre-arranged times, collecting from house to home, or community bin collection (central bin).

- Collecting equipment's are being implemented near slums, hotels, restaurants, commercial areas and others.

- Collected rubbish from residential and other places shall be transferred to the community bin by hand-driven carts or other small vehicles.

- Stray animals shall not be allowed to move around trash storage facilities or at any other site in the city or municipality.

- Biomedical wastes and industrial wastes shall not be mixed with municipal solid wastes and such wastes shall follow the rules separately specified for the purpose (**Figure 3**).

Figure 3.
Dumping of municipal waste. (Source: Municipal Waste Management in India, Recycling of Municipal Solid Waste (techpanther.co.in)).

3. Types of municipal waste

1. Organic waste: Kitchen waste, waste from food preparation, vegetables, flowers, leaves, fruits, and marketplaces.

2. Combustibles: Paper, wood, dried leaves, packaging for relief items, etc. This category of waste entails those that are highly organic and have low moisture content [5].

3. Non-combustibles: Metal, tins, cans, bottles, stones, etc.

4. Toxic waste: Old medicines, paints, chemicals, bulbs, spray cans, fertilizer, and pesticide containers, batteries, and shoe polish.

5. Recyclables: Paper, glass, metals, plastics.

6. Ashes or Dust: Residue from fires used for cooking.

7. Construction waste: Rubble, roofing, broken concrete, etc.

8. Dead animals: Carcasses of dead livestock or other animals.

9. Bulky waste: Tree branches, tires, etc.

10. Soiled waste: Hospital waste such as cloth soiled with blood and other body fluids.

4. The solutions for waste management that are preferred

- Reduction and reuse at the source: Preventing waste creation at multiple phases, such as product design, manufacture, packing, usage, and reuse, is the most recommended approach to waste management. Preventing waste lowers the cost of processing, treating, and disposing of it. Leachate, air pollutants, and greenhouse gases are also decreased, all of which have a major negative influence on the environment.

- Waste recycling: Recovery of recyclable material resources and fortifying a waste recycling chain through segregation, collecting, and reprocessing to produce new goods is the next favored option.

- Composting waste: In accordance with FCO guidelines, agricultural production and soil health can be enhanced by composting the organic portion of waste.

- Waste disposal: At the bottom of the hierarchy, residual waste—ideally made up of inert waste—should be dumped in sanitary, lined landfills built in compliance with the 2016 SWM Regulations.

5. Seven-step approach

The government has initiated various steps to dispose of the MSW. Here are the seven-step approach to collect waste and dispose of them (**Figure 4**).

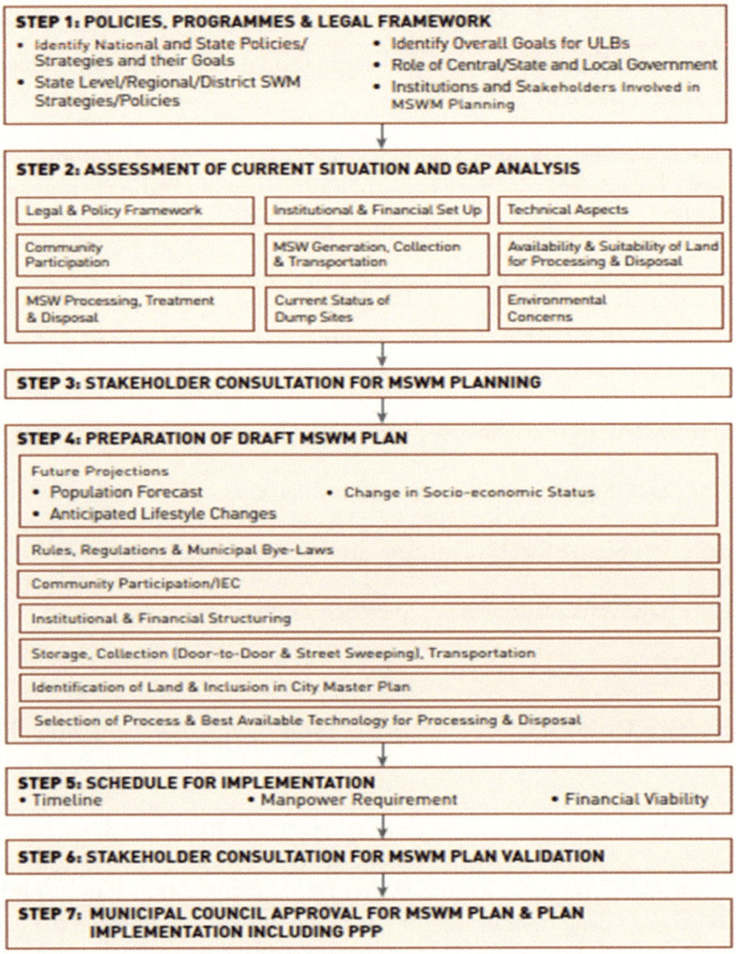

Figure 4.
Seven steps to collect and dispose of waste. (Source: [6]).

Step 1: Policies, programs, and legal framework.

Step 1 comprises a thorough examination and analysis of all applicable federal, state, and local MSWM-related laws, rules, policies, programs, and guidelines. To ensure that the MSWM action plan is developed within these frameworks, the ULB shall prepare a list of all mandatory and recommended actions as per the MSW (M&H) Rules, 2000; SWM Rules, 2016; the NUSP; the service level benchmarks (SLB) for MSWM service provision; the FCO, 2009, 2013; and all other pertinent policy guidance [6–8].

Step 2: Assessment of current situation and gap analysis.

Following national, state, and local laws, regulations, and MSWM plans controlling the ULBs, the municipal authority should critically evaluate the situation of SWM in the city. The evaluation needs to make evident any shortcomings or gaps that must be filled in order to comply with legal requirements. The assessment ought to concentrate on the characterization and quantity of waste. This is crucial since waste composition and quantity change greatly with urbanization and wealth.

Step 3: Stakeholder consultation for Municipal Solid Waste Management Planning.

According to national, state, and local level regulations, policies, and strategies for MSWM governing the ULBs, the municipal authority should conduct a critical assessment of the current state of SWM in the city. The assessment should highlight any gaps or deficiencies that must be filled to comply with legal requirements. It should also concentrate on waste quantification and characterization, which is crucial because waste quantities and composition vary greatly depending on urbanization and affluence. As a result of the numerous institutions and stakeholders involved in MSWM, it is crucial that the MSWM plan—which attempts to close the gaps or enhance the quality of service—be developed through consultation.

Step 4: Preparation of draft Municipal Solid Waste Management Plan.

The municipal authority should draft a short- and long-term MSWM plan after taking into account the gaps that have been identified, future population projections, waste generation rates, the quality and quantity of waste now and in the future (due to changing lifestyles and economic status), stakeholder input, financial constraints, and technical capabilities of the ULBs.

Step 5: Schedule for implementation.

The local government should outline the funding and institutional support requirements. The creation of an operational strategy should then follow as a crucial component of the MSWM plan. A strategy for implementation should be created, including the distribution of resources and providing deadlines. During this effort, public-private partnerships (PPPs) for the construction of infrastructure and the provision of services may also be thoroughly investigated.

Step 6: Stakeholder consultation for Municipal Solid Waste Management Plan validation.

Effective SWM service delivery is heavily reliant on the actions and customs of the community. As a result, before the MSWM strategy is finalized, stakeholders and the general public should be consulted.

Step 7: Municipal council approval for Municipal Solid Waste Management Plan and plan implementation including public-private partnership.

The elected body of the local authority must be presented with the completed MSWM plan to request approval and formally codify the plan. The financial plan and any institutional strengthening required for the implementation of these actions should be approved by the council in addition to being informed of the short- and long-term measures that need to be done. The ULB's MSWM plan should be communicated to the private operator or services to be outsourced (PPP model). Contracting models ought to be performance-based and open. To guarantee effective and long-lasting project implementation, both the PPP operator and the ULB should take responsibility for their respective responsibilities.

6. Waste collecting, segregating, and transportation

Until it is gathered for disposal by ULB employees or designated contractors, MSW should be kept in storage at the location of waste generation. Primary segregation, or the division of trash into distinct fractions, is a crucial process. The ULBs must prioritize the segregation of MSW and link it to the principal household waste collection program. Source segregation by waste generators will be useless unless the ULBs plan the primary collection of separated garbage.

Based on the trash's characteristics, the ULB's capabilities and resources, and additional framework factors (such as community customs, kabadi systems that are currently in place, and the amount of space that is available in streets and residential areas), the fractions into which the garbage must be precisely separated should be determined.

Garbage generators should, at the very least, separate their garbage into three fractions:

- wet (green container),
- dry (white container), and
- home hazardous waste.

This is known as basic segregation. The three-bin system is the name given to this. In addition to these wastes, sanitary garbage, waste from building and demolition, and horticultural waste should be collected and kept separately. To guarantee that sanitary waste is wrapped securely, collected, and sent to waste collectors separately, the wet part should ideally be utilized for composting. The dry waste should also be sent for recycling. Additionally, household hazardous trash needs to be gathered separately and dumped at the appropriate collection locations.

7. Segregated collection and transportation

One of the most important steps in SWM is the collection of separated municipal garbage at the source. The public's health and the beauty of towns and cities are impacted by inefficient waste collection services. Separate collection of dry and wet garbage improves the possibility of treating such waste economically and getting the most out of the recyclables added to the system (**Figure 5**).

7.1 Primary collections

Depending on the size of the city and the dominant waste management system, primary collection is the process of gathering waste from homes, markets, institutions, and other commercial establishments and transporting it to a storage depot, transfer station, or disposal site directly.

7.2 Secondary collections

Secondary collection entails gathering waste from transfer stations, waste storage depots, and community bins and delivering it to final disposal locations or waste

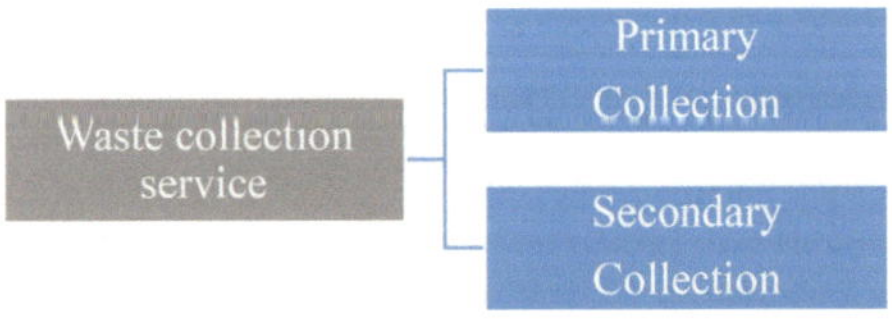

Figure 5.
Classification of waste collection service. (Source: [6]).

processing facilities. It is necessary to do primary collection in both big and small towns and cities. For sanitation workers to collect waste from community bins, secondary waste storage depots, or decentralized sorting centers and then transfer it to processing and disposal facilities, secondary collection systems are required in all cities and towns.

Waste spilling into streets and overflowing containers requires a well-coordinated primary and secondary collection and transportation system. To prevent handling garbage more than once, the transport vehicles should also be compatible with the waste storage depot's equipment design in addition to being able to transport separated waste.

They ought to be simple to maintain as well.

8. Street cleaning and drain cleaning

In India, municipal authorities have long provided the essential function of street cleaning to maintain clean and healthy urban environments. Prior until recently, the primary means of waste collection was street sweeping, and all household and commercial waste was dumped onto the streets or into street bins. Waste production has significantly decreased and its composition has changed in many metropolitan areas since door-to-door collection systems were implemented. Ideal street waste would contain dust, leaves, and some litter; but, until door-to-door collection starts working, there will also be significant amounts of recyclable and food waste collected during street sweeping. In India, manual sweeping is frequently done because of the country's many crowded streets and tiny roads.

There are many different tools and pieces of equipment available for both mechanical and manual sweeping. In order to enable the direct transfer of waste from a handcart to a container of a collection vehicle, municipal authorities must eliminate the need for additional waste handling procedures by converting conventional handcarts into containerized handcarts. Municipal administrations can accomplish considerable improvements in service quality and cost reductions by implementing effective strategies. In order to maximize staff productivity, ULBs should base the frequency of street cleaning on local conditions. Additionally, it is important to schedule street cleaning precisely to prevent issues with oncoming traffic, parked cars, and pedestrians.

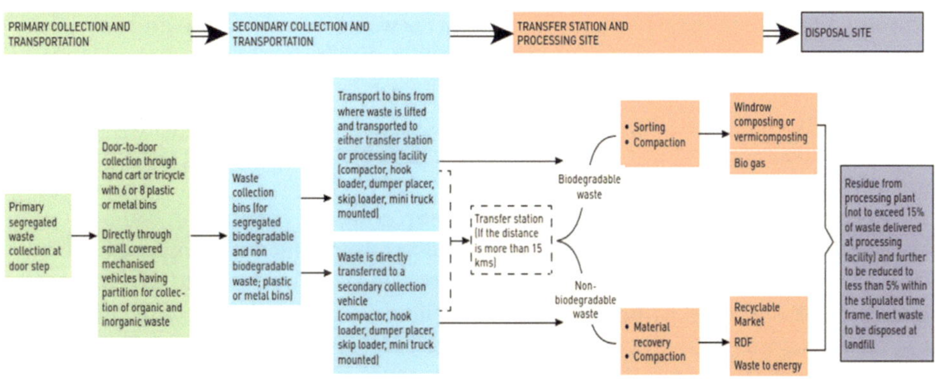

Figure 6.
Processing chart of waste collection and disposing. (Source: [6]).

Open surface drains by the side of the road are common in cities, and they must be cleaned regularly to allow storm or gray water to flow freely. SWM authorities should use education, public awareness campaigns, legal restrictions, and financial penalties to make sure that residents and sweepers do not put waste down the drain. Another way to stop this is to assign the same employees to clean the streets as well as the nearby surface drains, up to a depth of ninety centimeters (cm).

It is crucial to make sure that materials used for street sweeping and drain cleaning cannot be combined with waste that is collected from homes and businesses. This is because doing so can significantly reduce the options for treating and recycling home and commercial waste, as well as raise the expense of waste processing (**Figure 6**).

9. Processing technologies

9.1 Material recovery facility

Non-biodegradable or recyclable solid garbage that is collected from the doorway is placed in a material recovery facility (MRF), where it is processed and separated, and various recyclable waste components are recovered for sale. The setup of the MRF, which accepts combinations of waste fractions (recyclable or non-biodegradable), is dependent on several variables, including the kind, volume, and quality of incoming waste materials. Here, the material is essentially divided into streams of waste fractions (paper, plastic, bottles, packaging paper, etc.), and these streams are then sold to middlemen who provide the recycling companies with bulk material. In order to temporarily store sorted recyclables that can be made available to recyclers in bulk for increased resale value, MRFs also need extensive storage rooms. MRFs can be categorized as manual or mechanized, depending on the facility's level of automation and operational scale. Small-scale operations use manual MRFs, which are used for manual sorting. These units are usually run, owned, and controlled by the unorganized sector. Large-scale facilities have automated MRFs with complex machinery and systems that allow for the effective separation of vast amounts of material into various fractions.

9.2 Composting

The ISWM hierarchy lists resource recovery methods and composting as the third recommended waste management practice, after waste minimization and recycling systems. This ensures that garbage is treated adequately to enable further use of the item.

By biologically "digesting" the MSW through a controlled aerobic process, composting can be recycled for use as plant fertilizer, soil stability during cleanup, or soil amendment to help poor soils recover.

Depending on how feasible it is to implement, compost production can be done at two different levels: decentralized (household composting, bin composting, box composting, vermicomposting, in-vessel composting) or centralized (windrow composting, in-vessel composting, aerated static pile). Only separated organic matter can be composted, and both procedures necessitate extensive preprocessing. The compost that is produced needs to fulfill the quality standards outlined in the Fertilizer Control Orders (FCO) from 2009 and 2013. Before sizing the compost plants, it is important to determine the compost's market.

9.3 Waste to energy

Recovery of energy from MSW is proposed as a workable option when material recovery and composting from MSW are neither practicable nor acceptable because of local conditions or the nature of trash. High calorific value MSW fractions can be burned (thermal process) or their biodegradable fraction can be processed anaerobically (biomethanation). The energy produced in these processes—either as heat from burning or as biogas from methane—can be recycled or used to power machines. Waste-to-energy systems should become financially viable as a result of the sale of this energy. If the power tariff is not high enough to guarantee the plant's financial stability, the ULB might consider a tipping fee.

In order to achieve the necessary economies of scale, appropriate care should be taken to ensure a continuous and reliable supply of trash. It is also necessary to have enough environmental controls in place to adhere to strict incineration standards. There are numerous technological choices available; however, the choice of technology should be based on many factors, including the amount and quality of waste to be processed, the need for land, the financial ramifications, etc. These technologies can be used for varying waste quantities produced by various-sized urban local bodies. The Central Pollution Control Board (CPCB) or State Pollution Control Board (SPCB) may take into consideration recognized international emission control standards in cases where national legislation does not specify standards for emissions from various waste-to-energy technologies.

9.4 Incineration

As ash, flue gas, and heat are produced during the high temperatures and oxygen-containing combustion of garbage, the incinerator is a waste treatment method. That is possible with raw or garbage that has been treated very little, aside from the high calorific waste that has been separated.

The waste's composition, density, moisture content, and inert content affect its potential for energy production. It is possible to recover between 65 and 80 percent of the energy contained in biological matter as thermal energy. This is suitable for use in thermal applications. Only after appropriate material recovery and recycling mechanisms have been put in place, or in cases where there are no other viable processing options and there is a land availability issue, should incineration be taken into consideration. Garbage-to-energy plants should normally only be installed in cities that can supply at least 1000 tons of garbage per day (TPD). However, if plants are not run effectively and if suitable emission control measures are not implemented, incinerator plants have the potential to have a substantial negative impact on the environment through emissions and fly ash.

9.5 Biomethanation

The process of digesting biodegradable organic waste in an enclosed area with regulated temperature, moisture content, pH level, and other parameters is known as biomethanation. It occurs anaerobically, meaning without the presence of air or, more precisely, free oxygen. It is because of the high organic and moisture content of Indian MSW, this is seen to be one of the most technically feasible options. A steady supply of organic waste that is degradable and devoid of inert materials is necessary for biomethanation plants, as is a sustained market for the biogas they produce under the right economic circumstances. Depending on the waste input and implementation feasibility, biomethanation plants can operate at a centralized or distributed level (up to 5 TPD). The biomethanation plant's overall performance is heavily impacted by the input feed specification, and it needs segregated biodegradable MSW (such as hotel and restaurant waste). waste, market waste) for the best possible plant operation. From the perspective of efficiency, the feed material's uniformity is a crucial characteristic.

9.6 Refuse-derived fuel

The high calorific non-recyclable combustible component of processed MSW known as "refuse-derived fuel" (RDF) is utilized as an alternative fuel in industrial furnaces and boilers, or as a fuel source for steam and electricity generation boilers. Higher proportions of flammable elements than those found in the parent mixed MSW make up the makeup of RDF. Cement factories ought to co-process RDF if possible. Although it has not been tested in India yet, co-processing of RDF for the steel sector and electricity generation is also suggested. In other countries, co-processing RDF for power generation is a commonly used and technically validated waste management technique. However, there is limited information available regarding co-processing RDF in the steel industry.

9.7 Technologies under development

The treatment and processing methods for MSW that are commonly used are founded on extensive experiences across numerous nations. Regarding the suitability of emerging technologies for India, a few of these are being tested with assistance from the Indian government. It is best to wait for the test results.

These technologies should be treated as experimental technologies until they are well-established commercially. They should be managed as research and development (R&D) projects or through specifically created concession contracts. It might be considered a safe bet if new technology is successfully used for five years in the commercial world. India is experimenting with a number of technologies, including bioreactor landfills, gasification, and pyrolysis. Owing to the rising focus on waste to energy, even 300 TPD plants are said to be feasible with specific viability gap funding for plants based on gasification and pyrolysis; yet, none of these technologies has a track record of successful commercial implementation.

9.8 Construction and demolition waste management

Trash produced during any civil structure's construction, demolition, or remodeling is referred to as construction and demolition (C&D) trash. Rules for managing garbage from construction and demolition, updated in 2016, place a focus on the recycling and treatment of C&D waste.

C&D garbage includes nonhazardous insulating material, plastics, wallpaper, glass, metals (such as steel and iron), plumbing fixtures, concrete, bricks, tiles, stone, dirt, rubble, plaster, drywall, or gypsum board, wood, and asphalt, aluminum, etc.

Using their cozsir resources or by hiring private contractors, ULBs should arrange for the deployment of suitable containers (skips or other containers) and their removal at regular intervals or when they are filled operators. The gathered garbage needs to be moved to the proper location or locations for additional processing and disposal, either by hiring private operators or using their own resources. Under their purview, ULBs have an obligation to track and document the production of C&D waste.

The ULBs will prepare for the proper management of C&D waste, including processing facilities, and will further plan for the best possible use of the recycled goods after consulting with specialist institutions.

These organizations can also offer advice and support in this regard, suggesting methods to incorporate "deconstruction" activities from the beginning of the construction planning phase. In accordance with the updated regulations, municipal authorities must create bylaws and unique plans for the handling, processing, disposal, and storage of construction and demolition debris.

10. Difficulties in implementing MSW

Because the best and most appropriate techniques for waste collection and disposal are not being implemented, SWM currently has a low status in India. SWM training is lacking, and there are not many competent waste management specialists in the field. The existing SWM systems in India are likewise devoid of accountability [9]. In India, municipal authorities are in charge of managing municipal solid waste (MSW); yet, their budgets are insufficient to pay for the expenses of creating appropriate systems for trash collection, storage, treatment, and disposal. Effective solid waste management (SWM) in India is hampered by a lack of government finance regulations, waste collection/segregation, and strategic MSW strategies.

Low environmental awareness and motivation have hindered the implementation of innovative waste management technology in India. Public views toward waste provide a significant impediment to enhancing SWM in India.

11. Changes required to improve waste management in India

India's goal for waste management emphasizes using garbage as a valuable resource through extraction, recycling, recovery, and reuse. ULBs are responsible for waste management, with the Commissioner and Chairman holding direct accountability for system performance.

Waste management should be viewed as a necessary function in Indian society, with sustainable funding. To submit a case to a ULB for a well-funded waste management system, it is important to highlight the benefits of solid investment.

To improve waste management in India, a strong and independent authority is required to regulate the process. Without explicit regulation and enforcement, changes will not occur. Strong waste regulations can promote innovation. The waste management sector must incorporate attractive businesses that must meet ULB performance requirements or face financial fines applied when trash management services do not function properly. Finance for Waste Management Companies and infrastructure financing must be obtained from waste producers via a waste tax. An average price of one rupee per person every day would produce about 50,000 crores yearly, and thus effective garbage management in India may require sufficient money.

Accurate information on future waste amounts and characteristics is crucial for selecting effective waste management and treatment methods. Procurement of equipment and vehicles at the state level is important for primary and secondary collection, as well as proper monitoring of collection, transport, and disposal. Littering and garbage in streets is a big issue in India, with serious consequences for public health. Nagpur has implemented a road-sweeping method in which each employee sweeps a specific length of the road. UN-HABITAT picked the Centre for Development Communication's Swatchata Doot Aplya Dari (sanitary worker at your doorstep) project as a good practice example in 2007.

Effective waste management requires segregating garbage at the source to maximize value extraction and recycling. Separating dry (inorganic) and moist (biodegradable) trash offers significant benefits and should be the duty of waste producers. Long-term waste management planning necessitates visionary project development by ULBs, the business sector, and non-governmental organizations (NGOs). Define roles and responsibilities for delivering sustainable systems, and track progress through monitoring and evaluation. Experiences should be shared across India and among various social groups. There are several academic institutes, organizations, NGOs, and private sector enterprises working on a holistic approach to SWM, and future waste management in India would require considerable involvement of the informal sector across the system.

There is a need for training and capacity building at all levels. Indian schoolchildren should comprehend the significance of waste management, its impact on the environment and public health, and each individual's role in the system. This will create responsible citizens who see trash as a resource opportunity.

12. Conclusion

India is one of the world's developing countries, the population is growing and dumping of waste is increasing daily. In order to reduce the accumulation of waste, we can execute the seven-step plan that could help the government to protect the environment. As an individual, we are also responsible for disposing of waste in a sensible manner.

Author details

Mohana Priya Rajendren[1*] and Muthuminal Ramu[2]

1 Department of Food Technology, SNS College of Technology, Coimbatore, India

2 Department of Agriculture Engineering, SNS College of Technology, Coimbatore, India

*Address all correspondence to: mohana.r.ft@snsct.org

References

[1] Census of India. 2011 Ministry of Home Affairs, Government of India, New Delhi, India. Available from: http://censusindia.gov.in/ [Accessed: June 27, 2015]

[2] ISWA (International Solid Waste Association). 2012 Globalization and Waste Management Final Report from the ISWA Task Force. Available from: http://www.iswa.swachcoop.com.pdf

[3] Global Waste Generation - Statistics & Facts, Published by, Bruna Alves. 2023. Available from: https://www.statista.com/topics/4983/waste-generation-worldwide/

[4] Annual Report 2019-20 on implementation of Solid waste management rules. 2016. Available from: cpcb.nic.in

[5] Municipal Solid Wastes (Management and Handling) Rules. 2000. Available from: www.mpcb.gov.in

[6] Solid waste management manual. 2006. Available from: cpheeo.gov.in

[7] Sharma KD, Jain S. Overview of municipal solid waste generation, composition, and management in India. Journal of Environmental Engineering. 2019;**145**(3):04018143

[8] Gupta P, Sharma A, Bhardwaj LK. Solid waste management (SWM) and its effect on environment & human health. 2023

[9] Kumar S, Smith SR, Fowler G, Velis C, Kumar SJ, Arya S, et al. Challenges and opportunities associated with waste management in India. Royal Society Open Science. 2017;**4**(3):160764

Chapter 8

Reusing and Recycling Practices in Municipal Solid Waste Management in Kano Metropolis, Nigeria

Muhammad Saleh and Azizan Marzuki

Abstract

Kano Metropolis, currently Nigeria's second-largest city, has a population of 10 million, a significant increase from its pre-Saharan trade population. The city is primarily employed in farming, trade, and commerce, and is the most densely populated state in Northern Nigeria, with an average population density of 281 persons per square kilometer. The city generates 4-5 million tons of municipal solid waste weighing 1.12 kg/cap each day, with 0.5 million tons of unprocessed industrial waste annually, and poses a significant challenge to local handling due to its non-flow, evaporation, dispersal, and absorption into the environment. The state government and waste management firms have struggled to maintain clean streets, with over half of waste left uncollected, making the current system inadequate. Consequently, the current waste management practices in Kano Metropolis, Nigeria, with regard to MSW reuse and recycling are examined in this discussion. And then an overview is given on Kano Metropolis's current recycling and reuse strategies, as well as the kinds of items that can be separated from MSW and recycled again. Finally, the research study area and its surroundings are then encouraged to implement the strategies for enhancing recycling and reusable methods used in the management of MSW in Kano Metropolis, Nigeria.

Keywords: municipal solid waste, waste reusing, waste recovery, REMASAB, KNUPDA

1. Introduction

This study focusses on Kano metropolis, Nigeria's second-largest city, highlighting its population growth, economic activities in farming and commerce, and challenges with municipal solid waste (MSW). It notes significant municipal solid waste (MSW) generation, including industrial waste, with limited management success due to inefficient collection. The need for effective solid waste management (SWM) to enhance productivity, sanitation, and public health is to be emphasized. The study proposes examining current MSW reuse and recycling practices in Kano by suggesting strategies for improvement. Waste management in most developed and even in developing countries of which Nigeria is inclusive has increased in recent times due to increase in population, industrialization, urbanization, and globalization [1]. Whereas, based on its recent definition, solid waste management, in all ramifications, is simply a planned system aimed at effectively controlling the generation, storage, collection, transportation, processing, and final disposal of solid waste in landfill and or dumpsite [2]. This simply means that solid waste management is an important element for environmental protection, whose main purpose is to provide a hygienic, efficient, economic waste storage, collection, transportation, and treatment or disposal of solid waste without polluting the atmosphere, soil, or water system [3]. In another study, SWM refers to the generation, prevention, characterization, monitoring, treatment, handling, reuse, and residual disposition of solid wastes. In other words, solid waste management (SWM) is a precise name used for the collection, transportation, disposal, recycling, and monitoring/disposal of waste in other to avoid its adverse effects over human beings and the environment [4]. In fact, in Nigeria, all the solid wastes that are collected from the various localities usually end up in both designated dumpsites or uncontrolled landfills. For example, in Kano Metropolis the study area wastes dumped in open places, dumpsites, or landfills are usually set ablaze to reduce the volume occupying the space where the waste was earlier dumped. It then becomes obvious that these practices represent a basic problem in the country's methods of generating, collecting, storing, and treating solid wastes in Nigeria [5]. Although, it has been reported that what causes solid waste problem is not even the volume of waste produced but the degree of effectiveness the solid waste management processes it has undergone [6].

Therefore, using a cleaner technology that includes promoting waste recycling and reusable materials realized from waste and using suitable treatment for generated waste and adequate waste final disposal are some of the ways SWM problem can be resolved [7]. Environmental planning and management are becoming more well known. Nigeria has likewise acknowledged this requirement. Sadly, Nigeria has not yet created and put into effect environmental protection laws that would effectively address the solid waste management issues mentioned above [8]. Despite being the second largest and most industrialized city in Nigeria, Kano has regrettably earned a reputation as one of the dirtiest due to the abundance of rubbish that litters the city's landscape. Over the course of its history, Kano Metropolis has seen a massive increase in the generation of municipal solid waste (MSW). This is mostly due to the region's rapid population expansion as well as its economic and industrial development in comparison with other parts of the country, as it was previously indicated [9]. As such this waste management problem can be directly linked to its ever-increasing population, likewise, this problem is not different from those of other developing countries around the world, whereby it is reported by many researchers believes that management of (MSWM) has become a major environmental issue as shown in other SWM report study across Sub-Saharan African countries [10, 11].

According to **Table 1**, the Federal Ministry of Environment Nigeria's (2021) report and the Ministry of Housing and Local Government's (2015) report, for instance, demonstrate that Kano Metropolis, with its considerable industrial and social activity levels, has a high rate of solid waste generation. This table indicated that Kano

Metropolis alone has an estimated SW generation in 2020 of almost 3 million tons. This is almost the highest in the whole of the northern part of the country. Similarly, another report by Federal Ministry of Environment Nigeria (2021) has also estimated that the more than 8 million inhabitants of the state produce a total of almost 3 million tons of MSW in 2020. And in other to control the above wastes generation phenomena, then it is advisable to hopefully recycle and as well as reuse some the items that were being separated from this MSW [9]. And for long Kano Metropolis recycling and reusing activities in solid waste processing has been a source of concern first, to the government agency in charge of solid waste disposal and management (REMASAB) and secondly to residents (i.e., generators) residing in Kano Metropolis, are the major cause of this problem as it was observed during recognizance survey of the study area is mainly due to sharp practices in SWM from both the generators of this SW and the staff of the REMASAB whose services are not too frequent [12]. In fact, the Management of Urban MSW is a big problem in most cities of the world due to the lack of sufficient equipment to collect these wastes [11]. In fact, the rate at which waste is generated is so great that a refuse dump site that has been emptied earlier in the day could be replenished with an equivalent amount of waste the next morning, giving the false appearance that the site was never been cleared the day before [9]. In fact, these circumstances are not different from that of Kano Metropolis, Nigeria.

According to [5] there are five major aspects of MSW management, and accordingly these are: storage, collection, transportation, treatment, and disposal. This aspects or process must be followed to the latter in other to resolve some of these MSWM problems earlier mention above. Waste storage is the process of using the right equipment to store garbage at the point of generation until REMASAB collects it. The processes of collection and transportation entail gathering solid waste at the point of storage and moving it to a treatment facility or final disposal location. In contrast, treatment and disposal relate to a SWM approach that involves processing the material to recover resources that can be recycled, reused, or made unusable. The removal of the waste without endangering people or the environment is the last step in the process. Most final disposal can be accomplished by landfilling, even though this last process is always discouraging, since such disposal method is not sustainable in the last long [14, 15]. However, recycling and reusing process in SWM is reported to be a sustained method that can be adopted by all sundry in other to achieved

States	Estimated population 2010	SW generated 2010	Population 2020	Estimated SW generated 2020 ('000)	Average growth rate (population %)	Average growth rate (solid waste %)
Kogi	1,314,043	89,600	2,473,490	600,590	1.78	1.58
Kebbi	1,256,541	130,200	3,440,050	845,430	1.90	2.50
Kaduna	2,113,503	149,440	4,009,366	1,154,900	3.56	2.86
Kano	2,701,288	270,360	8, 206,892	2,980,660	4.55	3.95

Source: Refs. [12, 13].

Table 1.
Solid waste generation in Nigeria by states (in tons).

sustainable solid waste management in our urban cities. In fact, this method has been fully adopted by all developed country of the world, and this moment is achieving good results in MSWM [16].

2. Waste reusing and recycling practices

Waste handling, separation, storage, and processing at the source should involve segregation of waste at the source as this is expected to facilitate waste reuse and recycling and enhance waste reduction as well [17]. It was also on record that a community's willingness to separate waste is a significant predictor of waste reduction. Waste reduction, reuse or recycling, does not only reduce management cost, but it also mitigates environmental pollution and its impacts on public health [5]. For those in poverty who cannot afford to purchase new items, reuse can be extremely beneficial. These could consist of apparel, supplies for construction, and office machinery. Reuse focuses on gathering and distributing reusable items that can also benefit the community by helping with job-training programs and general training for young people, the disabled, and those without jobs for extended periods of time [1]. Finding new uses for previously wasted materials (i.e., wood, glass, plastics, metals, textiles, cardboard, paper, leather, rubber, and ceramics) as well as in the subsequent creation of useable commodities are further aspects of recycling wastes [18]. To put it another way, recycling solid waste is acknowledged as a workable way to clear the congested environment caused by the above aforementioned goods.

Reuse and recycling have also drawn more attention as a way to safeguard the environment since they provide one of the most practical answers to managing solid waste from an ecological and financial standpoint [4, 19]. In the waste management hierarchy, solid waste recycling and reuse come in the second and third place, respectively. The technologies of reduction, reuse, recycling, composting, incineration, and landfilling are extensively utilized in industrialized nations [20]. However, there are less negative effects on the economy overall when resources are reused as opposed to being used to create new products from raw materials, particularly if reuse leads to a decrease in the import of products and raw materials [1, 21]. For many people, reusing is an affordable option to get the things they require. Purchasing a used item as opposed to a new one is usually less expensive. Reusing reduces or eliminates the need for energy, water, and other resources, and it is unlikely to pollute the environment [15]. Reusing also has the added benefit of preventing environmental harm that would have resulted from discarding rather than reusing the item. On the other hand, using raw materials to manufacture a product (and, to a lesser extent, recycling them) results in resource consumption, pollution, and the generation of waste. According to early research, recycling was primarily industrial and influenced by financial concerns to lower production costs; however, today's focus on recycling is a means of conserving finite resources and minimizing waste from our environment [1]. In the study area, recycling programs and practices are still in their infancy. However, a limited quantity of recyclables, including papers, plastics, cans, and bottles, are stored in houses and then later sold to scavengers. In fact, most recycling facilities in Kano Metropolis are not legally recognized, the government should identify and keep an eye on them to ensure that environmental laws and regulations are being followed [22]. However, the activities of scavengers can have a great impact on the economy and waste management if properly undertaken [5]. But if the scavengers are properly organized, enlightened, and provided with the necessary economic and institutional support, it will eventually result in a sustainable waste practice because it will eventually become self-financing from the fees that will be collected from users of their service [8]. At present, however, their contributions are limited by the absence of government policy to encourage reuse and recycling. It has also been discovered that recycling previously used materials helps to mitigate the negative effects of rising waste volumes. People are involved in source separation, storage, collection, recycling, and disposal even though it has been reported that more than half of all solid waste is recyclable [23]. It is interesting to note that a significant amount of recyclable waste is thrown into the waste stream. However, the disposal of urban solid waste is becoming a bigger worry for the waste management industry in the developing nation due to a lack of public engagement in recycling programs and an apparent lack of finance for garbage transfer [21].

Recycling is a form of solid waste management (SWM) that involves the reprocessing of discarded materials, and recycling of previously used materials has also been found to alleviate deleterious impacts of increasing the amount of waste. More than half of solid waste is reportedly recyclable, but it is interesting to note that a significant portion of recyclable waste ends up in garbage [9]. As a result, people are involved in many aspects of solid waste management, including waste generation, source separation, storage, collection, recycling, and disposal. Paper recycling will lessen the need to cut down trees. Reusing metals can help cut down on mining operations [21]. In other words, recycling is the process of making something new out of waste materials including paper, bottles, broken glass, scrap metal, and other plastics. Furthermore, one waste management approach that can guarantee sustainability in any solid waste management plan is recycling [21].

Recycling is another waste management approach that can be utilized to stop or prevent materials that are thrown away from the waste stream. Waste segregation/sorting, processing, manufacture into new products, and other related tasks come before it. One technique for managing solid waste that can guarantee sustainability is recycling, which turns garbage from a liability to an advantage [3]. Also, waste is recycled casually by scavengers who had initially purchased these items or valuables from individuals, this individual goes a step further in the recycling process by visiting both lawful and unlawful dumpsites looking for materials that can be reutilized (i.e., reusing) and or recycled [5]. These goods include old car parts, an old refrigerator, an old drum, an old gas stove, an old office wooden door, old pieces of wood, aluminum sheets, old zinc/roofing sheets, and cans and tins of milk, tomatoes, and cooking oil. They produce a wide range of home goods that include: poultry products, building attachments, decomposing items for local manure production, and other materials that can be reused both inside and outside of their neighborhood or community [10]. The composition and properties of MSW in Kano Metropolis are given in **Table 2**.

3. Problems of MSW management in Kano Metropolis

From the above ongoing discussion, it was previously mentioned that the eight (8) local government areas that make up Kano Metropolis lack proper systems for SWM or equipment for the proper collection and management of waste. And this has the effect of causing the majority of Kano residents, who were seen in the study area

Components	Mass (%)	Properties	Composition	Mode of disposal	Potential uses
Paper/Cartons/ Cardboards	10	Combustible biodegradable Compactable and recyclable	10.50	Recycling/reusing	All paper products, e.g. paper roll
Plastics	4.6	Combustible/recyclable/biodegradable	9.4	Recycling/reuse	All plastics products
Textiles/Old Matrass	7.0	Combustible/biodegradable	4.2	Open burning/reuse	Local compost
Glass/Bottles	4.0	Rigid but can be recyclables	3.6	Recycling/reuse	Storage of produced
Metal/Aluminium	4.0	Recyclables/Reuse	5.20	Recycling/reuse	Fabrication of metal items
Miscellaneous	2.0	Non-combustible	20.34	Dumping	None
Dust/Ashes	0.6	Non-combustible	7.45	Dumping	None
Vegetables/Woods	59	Combustible	10.12	Dumping	Compost
Bones,	0.8	Non-combustible	8.04	Dumping	Feed mills
Leather	5.0	Biodegradable/reuse	7.81	Biodegradable	Reuse for leather products, e.g. shoes, bags, etc.
Inert/others	3.0	Non-combustible	13.34	Dumping	None
Total	100		100		

Note: Estimation is only based on the population and solid waste generated (2010 and 2020).
Source: Refs. [12, 13]. Federal Ministry of Environment Nigeria (2021); Ministry of Housing and Local Government (2015).

Table 2.
Composition and properties of MSW in Kano Metropolis.

paying fees to the unregistered private refuse/waste collectors to collect their domestic waste for disposal, to now typically dump their waste on the streets, in drainage areas, around street light poles areas, beside embankments, along secondary roads, and on the streets. And as a result, these practices have defaced the city landscapes, by causing air pollution, partially obstructed roads, and pedestrian pathways with refuse on them as it observed in the study area. Another problem of SWM is lack of adequate waste collection equipment's and vehicles for waste transportation; in fact, collection and transportation of solid waste requires adequate machinery and equipment. At the moment MSW is currently carried out over long distances, where MSW is transported to the four main designated treatment area or dumpsites.

The Refuse Management and Sanitation Board (REMASAB) and the Kano Road Traffic Authority (KAROTA) have been commendable in their efforts to keep the streets clean and highway free flow of traffic, but the lack of adequate equipment and waste collection vehicles has made their job a kind of a herculean task. Also, the indiscriminate dumping of waste has been reported to have defaced the city's landscape, especially for visitors coming into the city for the first. In fact, residents of Kano Metropolis dump their waste wherever it is convenient for them because there are no adequate designated waste collection points in either the residential or commercial areas. This is known to have created a collection problem, as most areas of the city are densely populated and improperly planned; most especially the core areas of Kano city, therefore, neither the REMASAB nor the Kano State Urban Planning and Development Authority (KNUPDA) officials/workers can reach these areas for wastes collection by giving it the desired planning proposal for sustainable waste disposal. And finally another problem at stake is the lack of continuity and implementation of government stronger legislative policies with those that are militating against the effective management of municipal solid waste in Kano Metropolis and Nigeria at large.

4. Current MSW management practices in Kano Metropolis

In Kano Metropolis, the main governmental agencies that have been entrusted with the responsibility of keeping the environment clean are the REMASAB, KNUPDA, State Ministry of Environment (SME), the Local Government Authority (LGAs), and the Ministry of Land and Physical Planning (MLPP). The government agency in charge of waste disposal usually transport waste into the four main waste treatment/disposal centre in other to undertake treatment of waste by mass burning (i.e. open incineration, which at the moment pollutes the environment). The state has four main disposal/treatment centres that are in full utilization, although there are other privately owned disposal centre scattered all over the Kano Metropolis, and they are also visited by waste scavengers in other to carry out waste separation for the purpose of acquiring resource for reuse and recycling. These four main treatment centres have no recycling machine, but a caterpillar stationed permanently for shifting the waste into the burrow pit. The process is simple; the waste is collected by the various agencies, (i.e. REMASAB, private waste contractors, waste pushers, etc.) who then transport them to the designated landfill/dumpsites, and from time to time these wastes are either separated manually by scavengers who source for reuse and or recyclables items or they are openly burnt in other to reduce their volume. For example, the discharge of hazardous gases and particulate matter into the atmosphere and harmful gases as a result of this action usually pollutes the environment; as such residents who live close to this open dumpsite should take particular note of this.

Although Kano Metropolis lacks an integrated waste management plan, it is clear that the current waste disposal systems have limits [9]. Open burning releases residuals into the atmosphere that must be disposed of in landfills [5]. Open waste burning produces ash, metal, and non-combustibles. Whereas, waste disposal is the last function or final step of solid waste management. The majority of waste disposals in Nigeria are open dump pit, a concept of burrow open-air pits. This concept is cheap, but fantastically rough on the environment. Improvements can be made on how some of the dump sites are operated, depending on the effectiveness of the managers in charge of the open dumpsites [10].

Therefore, just like in China [24]. Kano must now adopt programs that encourage complete waste reduction, reuse, and recycling of recyclables material in order to address the solid waste issues that were previously noted above. Anything that may be reused or put to better use should not be thrown away, especially when there is an urgent need for it elsewhere. [1] Whereas, the study of waste scavenge has revealed that most of the secondary raw materials that were scavenged from wastes are recycled by industries in the state, most especially plastic and metal/aluminum materials are recycled to produce new products. Other items from the waste dump are separated from the waste for reusing, if such waste practice is allowed to continue at a very higher rate, it is expected that there will be a reduction in the amount of waste earmarked for disposal as well as reduction in the energy associated costs by industries during production, and this is because recycling provides easily obtainable manufacturing feedstock [25, 26]. In Kano Metropolis, the cart pushers are among the groups of solid waste stakeholders engaged in waste reduction in the form of informal waste management. They are involved in house-to-house waste collection, transportation, and recovery of waste materials. Their activity usually concentrates on collecting a variety of metal scraps by cart pushing from either the waste dumpsites or to purchase these items from residential households or from ongoing building projects, where they buy some building metal items that will later be sold [22]. Other items they collect include old car batteries, metal vehicle spare parts, bottles, plastics, rubber, tires, etc.

5. Current reusing and recycling practices of MSW in Kano Metropolis

Waste recycling is an interesting approach towards achieving an efficient, integrated manner of managing municipal solid waste, and it also reduces the amount of household solid waste to be collected, transported, and disposed [18]. However, MSW recycling is restricted to a well segregated and clean high value materials [21]. The current emphasis on recycling is a way to reduce waste in the environment through on-source separation of recyclable items of MSW, which will require the preservation of scarce resources. Early studies have revealed that recycling in the past was mostly industrial and was based on financial considerations to reduce production costs [1]. Though only a small quantity of recyclables, including cans, plastics, bottles, and papers, are kept in houses and then sold to scavengers who then sales them to recycling companies since the study area still lacks a full-time recycling program. In fact, most recycling facilities in Kano Metropolis are not legally recognized, and the government should therefore identify this recycling facilities and as well as keep an eye on them so as to ensure that environmental laws and regulations in this regard are being followed to the latter [1, 14]. However, the activities of scavengers can have a great impact on the economy and waste management if properly undertaken [23]. This is because most of the recyclables waste resource used in recycling activities as observed in the study area are usually purchased from home of waste generators or separated by scavengers at the dumpsite and then sold to recycling industries or companies or Junk shop for further processing, whereas the reused waste resource also undergoes the same process as mentioned above [9].

Therefore, if the scavengers are properly organized, enlightened, and provided with the necessary economic and institutional support, it will eventually result in a sustainable operation because it will eventually become an economic venture and a self-financing from the fees that will be collected from users of their service [8]. At present, however, their contributions as observed in the study area are limited by the absence of government policy to encourage reuse and recycling activities. Recycling of previously used materials has also been found to alleviate deleterious impacts of increasing amounts of waste. There are typically 1–25 scavengers who sort recyclable and reusable materials at each of the four designated waste disposal sites in Kano Metropolis; however, the number of scavengers at each disposal site/dumpsite also depends on how large the solid waste dumpsite is. Scavengers were observed during reconnaissance survey to sort domestic recyclable wastes at disposal sites and refuse bins [9]. Despite reports and observations that more than half of all solid waste in the study area are recyclable resources, it is interesting to note that separated items from wastes either at the source of generation or at landfill sites/dumpsite by scavengers are sold to the junkshop owners [9]. This usually takes place after the materials are gathered, and then, they go through a stage of intermediate processing, like washing the items that can be reused and drying the items that can be recycled [8]. The materials that can be recycled and reclaimed are then sold to waste dealers, either at the junk shop or at the disposal site. **Table 3** shows the lists of the materials that are typically recycled or reused.

Table 3 above shows that, in Kano Metropolis, four main items are classified to be either recycled or reused, this includes plastics, metals glass, and paper, respectively; however, the majority of the secondary raw materials that are scavenged or purchased from dumpsites and are displayed in **Table 3** are not recycled by industries in Kano Metropolis, according to this research. This is because the majority of recycled industries also do not actively promote take-back recycling, which is a practice that is carried out

Recycled materials	Reused materials	Acquired at dumpsite	Acquired at the junkshops
Metals	Old shoes/textile	Yes	No
Copper	Polytene bags	Yes	No
Aluminum	Leather bags/box	Yes	No
Lead	Electrical components Wires	No	Yes
Blast	Old matrass	No	Yes
Paper	Tins/containers	Yes	No
Cartons	Kitchen utensils	Yes	Yes
Cardboards	Spare parts	No	Yes
Plastic bottles	Used tires	No	Yes
Glass	Oil jerry cans	No	Yes
Plastic wares	Carpets	Yes	Yes
White/color paper	Plastics containers	Yes	Yes
Plastic (PVC)	Woods/furniture	No	Yes

Source: (Field Survey, 2023).

Table 3.
Reclaimed of recycled and reused wastes items in Kano Metropolis, Nigeria.

Figure 1.
Workers separating and recovering items at a burrow pit waste dumpsite in Kano Metropolis.

in the majority of developed countries [27], and this should be discourage by the solid waste managers. It was also observed that scavenged or purchased metals sold have the highest selling price in the dumpsite or junkshop. However, it was observed that the price of metals varies depending on type, quality, and fluctuation of supply and demand of these metals. The scavengers were seen separating the materials further and reselling them to consumers, as well as supplying them for appropriate processing/remoulding to mills and factories. Thereafter, the processed material continues to proceed through the cycle. It is estimated that approximately 5–8% of MSW are recycled through this type of processing [17]. **Figure 1** shows scavengers separating and weighing waste resource items at a Burrow Pit Waste Dumpsite in Kano Metropolis.

6. Factors hindering solid waste hierarchy

Effective solid waste management is necessary for Nigeria to achieve sustainable urban development. A more effective strategy for preventive environmental

Figure 2.
Shows how municipal solid waste is being recovered and or recycled.

education must be used, and solid waste management hierarchy (SWMH) is an important option. Environmental education and public involvement opportunities raise awareness of the need of environmental protections for long-term solid waste management. A deteriorated environment due to lack of sustainable waste management is unsustainable for further expansion and has detrimental effects on a country's overall development [24]. In particular countries, a significant amount of SWMH takes place prior to the waste being dumped in a landfill. Rubbish pickers and scavengers gather materials from rubbish bins, waste collectors select materials that can be sold as they load their trucks, and scrap dealers buy directly from homes and businesses. According to [4], the analysis reveals that the local government prioritizes solid waste management above waste management hierarchy. On the other hand, since solid waste management hierarchy and solid waste management are two of the main theme of sustainable solid waste management, even though a prosperous nation should also have both. The research study area lacks official infrastructure for the storage, collection, separation of recyclable items, and disposal of solid waste, according to observations as well. Some families do not have adequate storage bins available [22]. Moreover, recycling bins that would aid in the separation of different waste categories are not given to families. Instead, every piece of garbage is placed into a single waste disposal receptacle that is located at the front of each home. Some families do not have adequate storage bins available [22]. Furthermore, families are not provided with recycling containers to facilitate the segregation of various waste categories. Rather, every household has a single waste disposal bin at the front where all wastes are piled together for collection by waste collectors (**Figure 2**).

7. Solutions towards addressing MSW management problems in Kano Metropolis-Nigeria

Basically in order to have a sustainable and efficient MSW management system in our cities through the use of recycling and reusing of resources separated from MSW, the following three aspects need to be taken into consideration, and these include: waste managers should target waste reductions at source of generation, technological intervention efforts, and institutional and regulatory reforms.

7.1 Waste managers should target waste reductions at source of generation

The effective management of MSW in Kano Metropolis Nigeria can be achieved through the development and implementation of laws, policies, and strategies to residents to enable them to reduce their waste generation at source through education and by setting mandatory standards that will penalize any defaulter. This law makes business owners to be responsible for the waste they generate. For example, such policies are mostly implemented in developed countries of Britain and the United States, and a regulation on recycling and reuse of packaging waste needs to be enacted. In United Kingdom, Local Municipal Governments area is mandated to collect recyclables items like plastic and paper packaging waste separately from other household wastes that are also separated for reusing. This law allows that producers and retailers to recycle the packaging wastes they produce; by this business owners are made responsible for the waste they generate; this practice is implemented by a designated privately owned recycling and reused companies, who are expected to after good negotiation by producers and retailers pays the company. This is similar to the practices in Japan and other developed countries of the world, in other words, the laws mandate waste managers the responsibility of allowing companies in reusing and recycling waste resources for production purposes, thereafter, the separated non-recyclables waste will now be disposed in landfill, and this has in fact, makes up the dual system in most of the SWM in our cities, and Kano Metropolis is not an exception.

7.2 Technological intervention efforts

The most effective technological interventions are those that support the environmental perspective that there are only resources and not waste. Waste will not exist due to this technological intervention because it will not be produced, and if it is, it will be separated into recyclable materials that can be used as resources again. This will evolve into an environment known as "industrial ecology," where waste from one area of the system becomes a resource for another, and our industrial society operates more like an ecological system [20, 28]. Integrated waste management (IWM), which is best described as a collection of management options that include: reusing, source reduction, recycling, composting, landfilling, and incineration, will be made possible by this technology intervention. For example, reusing involves separating waste material for reuse, but the availability of markets or junkshops along with favourable price structures for recyclable and reused materials, may encourage the continued production of waste items, whereas ignoring such options could eliminate wastes generation [8]. Furthermore, a product or material is recycled when it is reintroduced into the original production process through technological intervention efforts and used there as a new product. As a result, the following are some fundamental actions that can be taken in recycling practices: (i) source separation of waste, collection, and transport of wastes; (ii) sorting and cleanup of wastes for reused; (iii) processing of marketable materials by industries; and (iv) marketing and sale of the reuse product. In fact, the interrelation of the above basic steps will exert a strong influence on the economics of the recyclables waste scheme in our cities [23].

Therefore, the above basic steps should be considered and practiced more efficiently. Waste recycling and reuse can help eliminate and as well minimize wastes in our cities. Additionally, the process of incineration makes it quite possible to reduce waste. The MSW stream of Kano Metropolis has a higher proportion of organic matter than inorganic matter at the point of disposal, and this is according to the

REMASAB data on waste disposal that is currently accessible [9]. Nevertheless, it has a low heating value for energy recovery by incineration and a high concentration of components that are commercially recyclable. In the near future, reducing waste by incineration is not a highly practical technology for Kano Metropolis due to insufficient infrastructure of human, mechanical, and institutional resources, in addition to its greater cost when compared to alternative MSW management solutions [20]. Through composting, biodegradable wastes which make up a substantial component of Kano Metropolis' municipal solid waste can be transformed into organic fertilizer or soil conditioners. Waste managers need to advocate for this choice in any government program pertaining to municipal solid waste management [22].

Despite the fact that landfills are used extensively throughout the world for waste management, Kano Metropolis landfill sites should be carefully chosen because this is an important part of the waste disposal process. Furthermore, if this is done incorrectly, the created leachate—which is a byproduct of infrequent collection of waste for disposal—will adversely damage the waste management system's overall efficiency [29–31]. (LFG) which is a powerful greenhouse gas, and thus, uncontrolled landfill gas migration can result in a transfer of harmful gases which can damage the global environment and negatively impact human health [32]. However, if recycling and reusing practices in SWM is properly designed and operated, landfill gas which is the product in SWMH provides a source of energy that can be used for several energy producing purposes and thereby generate financial benefit for the landfill where such is dump [7].

It is vital to consider the chemical composition of solid wastes while assessing various alternatives for processing and recovery. Proximate and ultimate analysis of the substrate from the landfill activity is vital to know from the landfill activity, more especially when examining waste to energy processes where garbage is used as fuel. It should also be noted that for commercial recovery of generated LFG, a landfill should receive about 200 tons of waste per day, and in fact, the landfill or dumpsite should be designed for a minimum capacity of 500,000 tons of waste and must also have a minimum filling height of 10 m [32]. Also, the waste earmarked for landfill technology should not have been deposited for more than 5–10 years before LFG recovery is attempted [19]. This technology was observed to be absent in the study area, as such the need for its introduction in SWM in the study area. This technology will also help in adding to the electricity national grid that is low at the moment in the study area as a result of its high demand by the urban residents in the study area.

7.3 Institutional and regulatory reforms

Virtually every State Government in Nigeria has expressed concerned regarding urban solid waste management; yet, no city in Nigeria is exempt from the strain caused by unregulated urban solid waste management. The issue has been that most of the adopted regulatory reforms are insufficient. The identification of refuse collection points in every city is very important for the effective and successful implementation of any waste management programs and policies. However, institutional and regulatory reforms like the public knowledge must be pursued vigorously. In order to educate its people about the need of maintaining a clean environment through the proper disposal of their waste at designated collection places, the government must make use of every means of communication resources at its disposal. This will prevent indiscriminate dumping of waste in places other than the designated collection points. The Kano State Government is required to create environmental policies and regulations that encourage the reuse and recycling of municipal solid waste (MSW) and to enforce adherence to these practices throughout the Kano Metropolis's waste-generating areas. Although, Kano Metropolis is not yet ready for the practice of energy recovery from incineration and as such it continues to landfill most of its MSW in the four main burrow pits. Kano Metropolis can, however, maximize the benefits from the MSWs by recovering LFG and generating energy from the gas and as well as separate items for reusing. One cannot emphasize the advantages of using landfill gas recovery, particularly for producing electricity to supplement the current insufficient supply from the national grid. Consequently, by recovering LFG and producing energy from the gas, MSWs can be utilized to their fullest potential.

8. Conclusion

In conclusion, this research report, therefore, now recommended that in view of the wearisomeness of urban solid waste management problem more especially in recycling and reusing practices, then a sustained cooperation should be developed among all key actors in SWM, and additionally, the establishment of waste/sanitation committees in state houses of assembly and the appointment of waste management advisers to state governors are both necessary. These legislative committees will specify proper recycling and reuse policies and guidelines that will direct the management of urban solid waste as well as consumer products that really produce waste, especially non-biodegradable ones. The government should also implement SWM rules which requires urban dwellers to separate their waste at the source through these legislative committees on solid waste management. This expected to pave way for proper recycling and reusing practices in the study area. Whereas, these key actors in SWM should include government, private waste companies and managers, and public health workers and inhabitants.

These key factors are expected to also implement an economic, sustainable, and reliable solid waste management practices such as the sustainable solid waste management hierarchy (SSWMH) and solid waste management system (SWMS) that consist of reuse, recycle, recover, reduction (i.e., 4Rs) and generation, separation at source, collection, processing, transfer, transportation, and disposal also known as SWMS, respectively. This would help waste managers and others stakeholders in the SWM sector towards realizing the potential resources that could be generated from the waste materials that has been recovered from waste or that can be reuse again [19]. In other words, solid waste management must be recognized as an important issue within the environment, and as such it must be linked to community participation, environmental education, poverty alleviation, and the creation of employment for all sundry in the community. As a consequence, its supervision and control must be held with the local authorities/waste managers who are in a better position to supervise this process. In fact, it is an issue that essentially require the involvement of private operators, especially small operators, who will not only create jobs and wealth in the local communities, but will invariably help in development of local technologies since equipment that is to be used must adapt to the local conditions [12]. This suggests that in order for every segment of the society to fully understand and appreciate the fundamentals of a high-quality environment, they must be appropriately informed and appropriately educated. If this occurs, managing urban solid waste becomes a shared responsibility, requiring everyone to get involved in other to make the study area's urban waste management successful.

Acknowledgements

Appreciation goes to Refuse Management and Sanitation Board (REMASAB) a Kano State Government Agency in charge of waste management, Ministry of Environment in collaboration with the Kano State Urban Planning and Development Authority (KNUPDA), and the State Ministry of Housing who help with the much-needed information that this study required to be successfully completed.

Conflict of interest

The authors declare no conflict of interest.

Acronyms and abbreviations

REMASAB	refuse management and sanitation board

Data and material availability

Information regarding this study is available by contacting the corresponding author.

Author details

Muhammad Saleh[1,*] and Azizan Marzuki[1]

1 School of Housing, Building and Planning, Universiti Sains Malaysia, Pulau Pinang, Malaysia

2 Bayero University, Kano, Nigeria

*Address all correspondence to: salehmuhammad@studentusm.my; plannersaleh2016@gmail.com

References

[1] Ibrahim MS, Saidu DAA, Abubakar AJ, Aliyu F, Adamu A. Waste scavenging as a tool for poverty reduction among unemployed youths in Bauchi Metropolis, Nigeria. African Journal of Management and Business Research. 2022;**1424**(1):136-156. Available from: https://publications.afropolitanjournals.com/index.php/ajmbr/article/view/

[2] Ike CC, Ezeibe CC, Anijiofor SC, Daud NNN. Solid waste management in Nigeria: Problems, prospects, and policies. 2018;**2009**:163-173

[3] Igbinomwanhia DI. Status of Waste Management. Vol. 13. London, UK: Intech; 2016. Available from: https://www.intechopen.com/books/advanced-biometric-technologies/liveness-detection-in-biometrics

[4] Salami HA, Adegite JO, Bademosi TT, Lawal SO, Olutayo OO, Olowosokedile O. A review on the current status of municipal solid waste management in Nigeria: Problems and solutions. Journal of Engineering Research and Reports. 2019;**3**(4):1-16. DOI: 10.9734/jerr/2018/v3i416884

[5] Babayemi and Ogundiran. Current Levels and Management of Solid Wastes in Nigeria. 2017. 29-53. doi:10.1002/tqem.21498

[6] Uwadiegwu BO. Strategies for effective urban solid waste management in Nigeria. European Scientific Journal. 2013;**9**(8):296-308

[7] Mukherjee C, Denney J, Mbonimpa EG, Slagley J, Bhowmik R. A review on municipal solid waste-to-energy trends in the USA. Renewable and Sustainable Energy Reviews. 2020;**119**(October 2019):109512. DOI: 10.1016/j.rser.2019.109512

[8] Nabegu AB, Mustapha A. Institutional constraints to municipal solid waste management in Kano Metropolis, Nigeria. International Journal of Innovative Environmental Studies Research. 2015;**3**(July-September):13-21

[9] Lambu IB. Waste or wealth' the cultural crux behind scavenging in urban Kano state, Nigeria. 2016;**3**(3):19-25

[10] Eeda AN. Integrated Sustainable Household Solid Waste Management Using Solid Waste Minimisation Approach in Shah Alam, Selangor (Vol. 4, Issue March). Malaysia: Faculty of Built Environment Universiti Teknologi Malaysia MARC; 2017

[11] Justice, Kofi Debrah DGV, Vidal DG, Dinis MAP. Raising awareness on solid waste management through formal education for sustainability: A developing countries. 2021;**6**(6):1-21

[12] Nabegu AB. An analysis of municipal solid waste in Kano Metropolis, Nigeria. Journal of Human Ecology. 2010;**31**(2):111-119. DOI: 10.1080/09709274.2010.11906301

[13] Umar M, Naibbi AI. Analysis and suitability modeling of solid waste disposal sites in Kano metropolis, Nigeria, Geocarto International. 2019. DOI: 10.1080/10106049.2019.1655796

[14] Alhassan KA. Environmental sanitation problem and solutions a case studies of Kano municipal local government area, Kano, Nigeria. 2015;**3**(7):241-251

[15] Balogun DO, Ibrahim A, Mshelia AN, Augustine A, Adgidzi JA, Boyi S.

Municipal Solid Waste Management Practices and Impact on the Environment in Nasarawa Local Government Area. World Wide Multidisplinary Research and Development. Research gate; 2020. pp. 15-22

[16] Bautista PR. Level of awareness and practices on solid waste management (SWM) among college students. Journal of Biodiversity and Environmental Sciences. 2019;**2019**(1):131-138. Available from: http://www.innspub.net

[17] Sharma KD, Jain S. Overview of municipal solid waste generation, composition, and management in India. Journal of Environmental Engineering. 2019;**145**(3):04018143. DOI: 10.1061/(asce)ee.1943-7870.0001490

[18] Ukala DC et al. A review of solid waste management practice in Nigeria. NIPES Journal of Science and Technology Research. 2020;**2**(3):169-174. pISSN-2682-5821, eISSN-2682-5821. 10.37933/nipes/2.3.2020.17 © 2020 NIPES Pub. All rights reserved

[19] Pratap V, Bombaywala S, Mandpe A, Khan SU. Solid waste treatment: Technological advancements and challenges. In: Soft Computing Techniques in Solid Waste and Wastewater Management. 2021. pp. 215-231. DOI: 10.1016/B978-0-12-824463-0.00014-8

[20] Karim MA, Corazzini B. The current status of MSW disposal and energy production: A brief review of waste incineration. MOJ Ecology and Environmental Science. 2019;**4**(1):33-37. DOI: 10.15406/mojes.2019.04.00129

[21] Kofoworola OF. Recovery and recycling practices in municipal solid waste management in Lagos, Nigeria. 2007;**27**:1139-1143. DOI: 10.1016/j.wasman.2006.05.006

[22] Winter M, Ujoh F. A review of institutional frameworks and financing arrangements for waste management in Nigerian cities. Urban Studies and Public Administration. 2020;**3**(2):21. DOI: 10.22158/uspa.v3n2p21

[23] Kumar A. Dumping yard-free society through sustainable solid waste management: A case study of Varanasi City. In: Waste Management and Resource Efficiency. 2019. pp. 409-421. DOI: 10.1007/978-981-10-7290-1

[24] Steuer B. Is China' s regulatory system on urban household waste collection effective? An evidence-based analysis on the evolution of formal rules and contravening informal practices. Journal of Chinese Governance. 2017;**2**(4):411-436. DOI: 10.1080/23812346.2017.1379166

[25] Permana AS, Towolioe S, Aziz NA, Ho CS. Sustainable solid waste management practices and perceived cleanliness in a low income city. Habitat International. 2015;**49**(November 2017):197-205. DOI: 10.1016/j.habitatint.2015.05.028

[26] Pujara Y, Pathak P, Sharma A, Govani J. Review on Indian municipal solid waste management practices for reduction of environmental impacts to achieve sustainable development goals. Journal of Environmental Management. 2019;**248**(April):109238. DOI: 10.1016/j.jenvman.2019.07.009

[27] Pires A, Martinho G, Rodrigues S, Gomes MI. Sustainable Solid Waste Collection and Management. 2018. doi:10.1007/978-3-319-93200-2

[28] Momodu NS, Dimuna KO, Dimuna JE. Mitigating the impact of solid wastes in urban centres in Nigeria. Journal of Human Ecology. 2011;**34**(2):125-133. DOI: 10.1080/09709274.2011.11906377

[29] Kumar A, Agrawal A. Current research in environmental sustainability recent trends in solid waste management status, challenges, and potential for the future Indian cities—A review. Current Research in Environmental Sustainability. 2020;**2**:100011. DOI: 10.1016/j.crsust.2020.100011

[30] Majeed A, Batool SA, Chaudhry MN. Informal waste management in the developing world: Economic contribution through integration with the formal sector informal waste management in the developing world: Economic contribution through integration with the formal sector. Waste and Biomass Valorization. 2017;**8**(3):679-694. DOI: 10.1007/s12649-016-9648-4

[31] Parvez N, Agrawal A, Kumar A. Solid waste management on a campus in a developing country: A study of the Indian institute of technology Roorkee. Recycling. 2019;**4**(3):1-22. DOI: 10.3390/recycling4030028

[32] Saleh HM, Koller M. Introductory chapter: Municipal solid waste. Environmental Development. 2019;**2/4**(3):1-8

Section 5

Waste Segregation and Source Management

Chapter 9

Significance of Source Segregation of Solid Waste in the Course of Solid Waste Management: An Overview of Varanasi City, India

Awadhesh Kumar and Narendra Kumar Rana

Abstract

Improving solid waste management (SWM) is a major challenge for governments globally and in India, particularly due to the lack of source segregation. If source segregation is not properly implemented, valuable materials that could be repurposed end up in landfills, putting a strain on the use of virgin raw materials. Several factors such as financial constraints, inadequate infrastructure, lax regulations, overcrowded urban areas, and the "not-in-my-backyard" (NIMBY) attitude hinder the implementation of integrated waste collection and treatment programs. Therefore, it is crucial to make a paradigm shift and view waste as a valuable resource rather than as worthless objects. This study aims to highlight the benefits of municipal solid waste (MSW) source segregation and explore the challenges and technical aspects of implementing it in Varanasi city. In 2022, Varanasi was divided into eight zones and one hundred wards. Using Yamane's formula, the sample size was determined from the universe of households. Four hundred and forty families in Varanasi City participated in the study from January to March 2024. The results indicate that many residents believe waste segregation is futile because the city collects and transports all waste together. However, education and infrastructure improvements could help to improve source separation.

Keywords: source segregation, waste material, landfills, NIMBY, paradigm shift

1. Introduction

The term "waste" often denotes an item that is disposable and of no use to the original user. Solid wastes (SW) are the stuffs that are generally discarded haphazardly close to the residential areas. Much of what would normally be thrown away can be recycled, reused, and even brought back to the full circle if the SW materials are precisely sorting at the place of origin, and it may also simplify the handling of waste materials. Besides that, additional strain on virgin or fresh items may be minimized, but efficient municipal solid waste (MSW) management is now widely recognized as one of the most important environmental issues facing the planet [1, 2]. Conditions for SWM improve as nations move from lower-income to middle-income and finally high-income levels; nevertheless, when cities and populations expand quickly, waste collection, dumping, and treatment become more difficult [3]. Research has shown that the amount of waste produced positively correlates with economic growth, population increase, urbanization, and industrialization of the nation [3, 4].

The enormous volume of solid waste puts additional strain on the infrastructure of cities. Simultaneously, several factors impede cities in enhancing waste management infrastructure, including financial and human resource constraints, inadequate institutional capacity, insufficient regulatory frameworks, limited enforcement, insufficient infrastructure, dearth of urban land for waste disposal, and societal resistance because of not-in-my-backyard (NIMBY) culture [5]. Ineffective and subpar SWM can result in public health issues or environmental issues from the uncontrolled release of methane and leachate [3]. Instead of merely burning or burying MSW, it is urgently needed to make a paradigm shift that views waste as a resource rather than as worthless and useless objects [5]. Source separation may be a good strategy to raise waste management standards and transform it into products or valuable resources [6], and innovative techniques for source separation must be implemented [7] to ensure that recyclable materials are diverted from reaching landfills.

Oduro-Kwarteng et al. [7] assert that the process of source separation for MSW involves the segregation of postconsumer materials at the point of generation so that they do not become part of the mixed waste streams [7]. Yang et al. [8] stated that source segregation involves the process of isolating various materials in MSW at their point of generation based on their unique characteristics, with the aim of facilitating subsequent treatment processes [8]. As instructed by the Government of India's SWM Rules in 2016, waste producers are required to categorize and store their waste into three separate groups: biodegradable, nonbiodegradable, and domestic hazardous waste. Furthermore, they must hand over the segregated waste to authorized waste collectors [9].

Oduro-Kwarteng et al. [7] highlight sorting excessively and assert that there are two types of waste sorting methods, mechanical sorting and source sorting, which are used to collect the organic fraction of municipal solid waste (OFMSW) [7]. The organic part of municipal solid waste that is source sorted is collected separately from markets, canteens, restaurants, and other establishments and is sorted by hand at the point of waste generation. At a materials recovery facility, mixed trash is separated into the mechanically sorted organic fraction of municipal solid waste (MS-OFMSW). This MS-OFMSW is contaminated and unfit for use as compost in agricultural settings. Kaseva et al. [10] and Bennagen et al. [11] declare that it is more advantageous to separate recyclables at their source since doing so results in cleaner and superior quality materials, as opposed to recovering mixed waste at a material recovery facility [10, 11]. Nevertheless, source separation is more prevalent in developed countries, whereas it is almost nonexistent in developing nations [7].

Furthermore, the collection and transportation of sorted waste items to the treatment facility or landfill site is the next task facing the authorities. Once collection services are delayed, wandering animals and fleas—especially when it comes to food waste—can cause significant annoyance [12]. The hearts of waste collection are the waste generation points, which include homes, businesses, educational institutions, market and public places, and so forth. Various sensor-based waste receptacles of different colors and caution could be useful for sorting and collecting rubbish from markets and public areas, while door-to-door waste collection is the most popular method that may be used to collect separated waste from households, businesses, and educational institutions. Sadly, in most of the urban areas of the developing countries including the research area, MSW is collected and dumped in a mixed form. If processing facilities are not interconnected, the challenge of intermingling discarded items can be improved by having collection trucks with different chambers for different MSW components or different vehicles for distinct MSW components [12].

Garbage collected from various sources is then finally assembled at transfer stations or secondary storage facilities (SSF) prior to being transported to landfills or treatment facilities. When it comes to final transit, most rubbish is prone to becoming mixed up. Waste from transfer stations or SSF should be transported in sufficiently large, clearly marked (such as green or blue) vehicles carrying specialist waste, such as biodegradable or wet waste, nonbiodegradable or dry waste, or hazardous garbage, to the processing units or landfill sites [12]. By doing this, mixing waste might be prevented. In addition, each of these vehicles needs to have an enabled GPS system in order to track their movements and avoid combining garbage at any point during transportation [12].

When presenting the waste management scenario of the research area, Varanasi city, the condition of waste segregation at the source has not yielded the expected outcome. Even though various waste collection methods are being executed, such as door-to-door collection, street sweeping, collection from marketplaces and public areas, and so on, the mixed waste is eventually transferred to the waste treatment plant as a whole. Therefore, source segregation becomes a futile endeavor. The current study aims to assess several aspects of the existing state of source segregation of waste as well as the MSW management scenario in Varanasi city. The proposed structure can be useful for drafting a source segregation plan for the city. Other Indian cities with similar garbage characteristics might potentially benefit.

2. Objective

The primary focus of the study is to propose the methods for waste separation at its point of origin. This includes households and public areas, where segregation is feasible. Additionally, the study will recommend the same method of collection and transportation for the segregated waste, in order to fulfill its intended purpose.

3. Study area

The research has been conducted in Varanasi, a city that is renowned for its historical significance and is considered one of the oldest continuously inhabited cities in the world. Located on the crescent-shaped left bank of the River Ganga, Varanasi is often referred to as a city of prayer (**Figure 1**). Spanning 6.5 km (4 miles) in length, the riverfront forms the eastern boundary of the city and is situated within the Indian state of Uttar Pradesh [13]. The city spans from 25° 14′ North 25° 23′ North latitude and from 82° 55′ East 83° 3′ East longitude. *Varunapar*, *Adampur*, *Kotwali*, *Dashashwamegh*, *Bhelupur*, *Rishi Mandav*, *Sarnath*, and *Ram Nagar* are the eight administrative zones that jointly constitute Varanasi City. These zones collectively comprise 100 wards, covering an area of 150.69 sq. km and housing 16,36,659 people in this great cultural and religious capital.

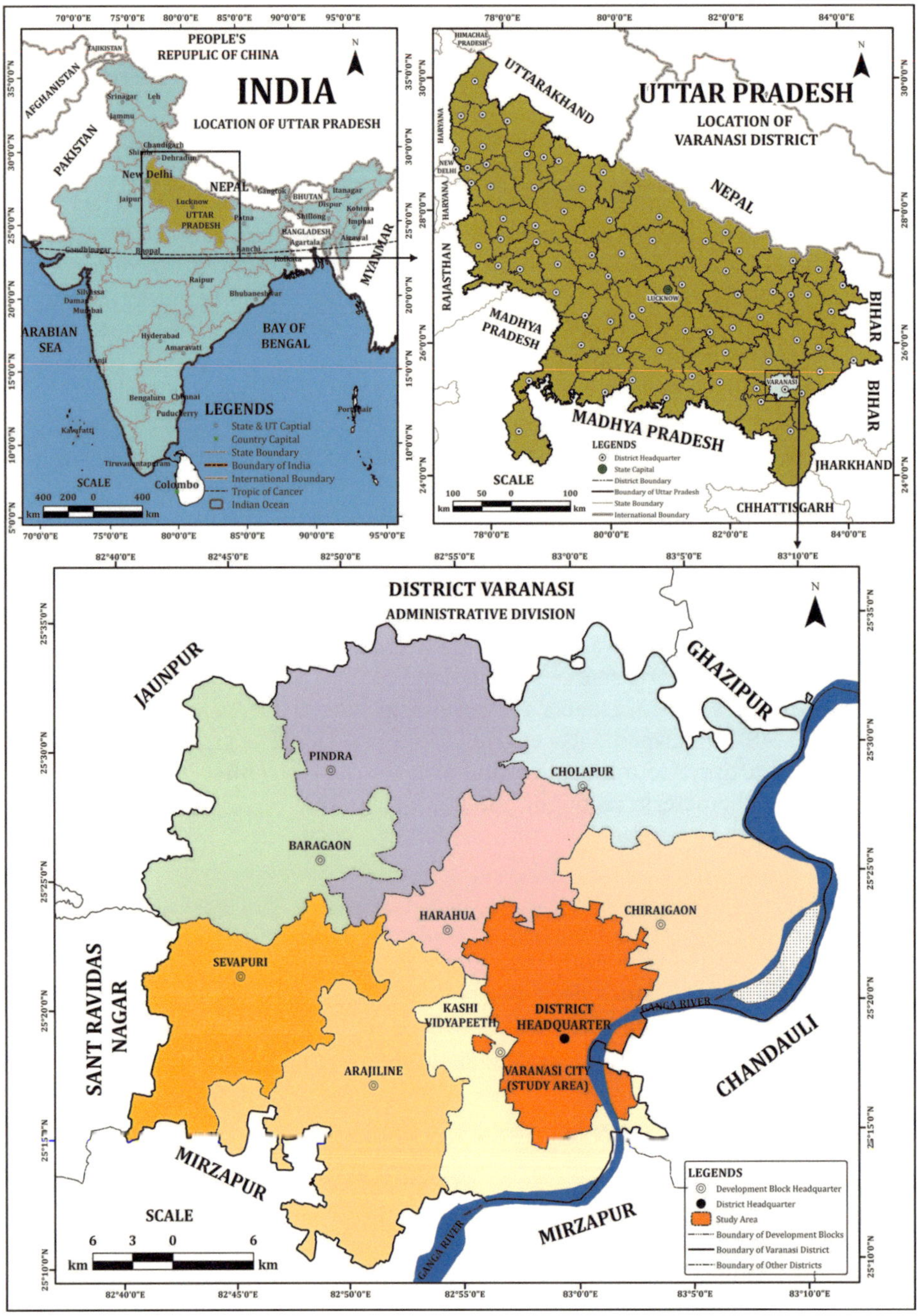

Figure 1.
Location map of the study area.

Aiming to become "Nirmal Kashi," (Clean Kashi) the city has enhanced scientific sustainable solid waste management in place as part of its Smart City Plan [14], but the research area in question is confronted with a multitude of challenges associated with high urban density and traffic congestion, both of which have a detrimental impact on the efficiency of solid waste collection [15].

4. Implemented methodology

This research employs a case study analysis approach due to the importance of the subject matter, and Varanasi City was selected as the focal point of the study. Moreover, an empirical inquiry utilizing a conventional content analysis approach was carried out in the study area from January to March 2024 to gather both qualitative and quantitative information. The objective was to investigate the impediments that hinder the implementation of source segregation of solid waste in the aforementioned research area.

In the year 2022, Varanasi City was divided into eight zones and one hundred wards, and the delimitation of the aforementioned wards was carried out based on a population of 16,36,659 [16]. As the census of 2021 was not available owing to COVID-19, the average number of individuals per household in 2011, which was 6.28, was utilized to calculate the number of households in 2022. The resulting figure was 2,60,606 households, which accommodated a total of 16,36,659 individuals within the study area. Employing Yamane's formula, the sample size was determined based on the household population. Adding 10% more to the total number of samples, as recommended by Barlett et al. [17], accounted for unresponsive or unreachable respondents. In total, 440 families from Varanasi City participated in this empirical investigation.

Using Yamane's formula, the sample size has been computed [18].

$$n = N/1 + N\,(e)^2$$

Where 'n' is the sample size, 'N' is the population size, and 'e' is the level of precision. Where N = 2,60,606; e = 0.05 or ±5; confidence level = 95%.

$$n = \frac{2{,}60{,}606}{1 + 2{,}60{,}606(0.05)^2}$$

$$n = \frac{2{,}60{,}606}{1 + 2{,}60{,}606 \text{ X } 0.0025}$$

$$n = \frac{2{,}60{,}606}{1 + 651.515} \quad n = \frac{2{,}60{,}606}{651.515}$$

$$n = 400$$

$$n = 400$$

According to Barlett et al., 10% of the sample size has been added to the total number of samples in order to compensate for respondents who were unable to get in touch with the survey, or some values missing, or have data with outliers [17]. Hence, the final sample size was 440.

$$n = 400 \text{ x} \frac{10}{100} = 40$$

$$n = 400 + 40$$

$$n = 440$$

4.1 Interview process

The interview process was conducted with a total of 440 respondents, with face-to-face interaction and the use of semi-structured schedule-based questions. The questions were designed to be both open-ended and close-ended, allowing for the collection of both qualitative and quantitative information for the purpose of analysis. A focus-group discussion was also steered with Varanasi Municipal Corporation (VMC) personnel who were involved in solid waste management, as well as waste management staff in the field.

Open-ended responses were recorded with the pre-consent of the respondents, assuring that the information would be used solely for the purpose of the study and thus avoiding any loss of information. The initial question asked was whether the respondent was familiar with the term “source segregation” and, if so, how they practiced waste segregation. The interview then proceeded with additional relevant questions. A personal observational study was also conducted to assess the severity of the problem. Rag pickers, an unofficial sector, were also questioned to emphasize their contribution to garbage segregation and management.

5. Review of existing literature

In the literature review, the focus has been placed on the challenges faced by individuals in segregating discarded items, as well as the collection and transportation issues experienced by workers involved in the handling and processing of solid waste. Studies from a diverse range of geographical locations, such as Iran [19, 20], Ghana [7, 21, 22], Shanghai, China [23], India [24, 25], Japan [26], and Ethiopia [27], have been included in this research.

In Ref. [25], the findings of Rathore and Sarma’s study highlight four primary factors that affect the inclination of urban residents in India to separate their waste at source are perceived advantage, availability of separation facilities, market characteristics, and social awareness. Moreover, the study indicates that attitude serves as a significant mediating variable for these factors. In Ref. [20], Babazadeh et al. pinpointed six key challenges in the implementation of household solid waste separation at its source: (a) limited engagement from citizens, (b) theft of garbage, (c) unequal distribution of resources, (d) weakness in the policymaking process, (e) underperformance of the private sector, and (f) absence of integrated management.

In Ref. [21], Alhassan et al. found in Ghana’s millennium city that source separation behavior is influenced by various factors, including household location, type of service provider, attitude, income, gender, and monetary incentives. In Ref. [19], according to the authors Babazadeh et al., there are four primary categories of hindrances that obstruct source separation, such as obstacles related to the collection system, the absence of individual responsibility, inadequate public awareness, and anticipation of receiving incentives for participation. In Ref. [24], Singh and Raj included a logistic regression model in their study and concluded that assistance with source segregation is the most effective approach for sustainable waste recycling in Patna, whereas in Ref. [7], Oduro-Kwarteng et al. reported that the success of source separation programs depends on education and economic incentives provided to waste generators.

In Ref. [22], Oduro-Appiah and Aggrey found that residents of Thema (Ghana) possess a strong determination to separate their waste into two categories, biodegradables and plastics. However, the main obstacles are finding bin storage place and making sure that the separated garbage is regularly collected. In Ref. [23], Zhang et al. observed in their study that although the respondents were environmentally conscious about separation, they only practiced minimal separation. Negative neighbor effects, confusion in classifying MSW and mixed transportation and disposal are the major limitations of source-separated MSW collection. In Ref. [26], Matsumoto discovered in his study that the recycling programs in Japan are customized based on the demographics of the population. In Ref. [27], Adefris et al. emphasize the importance of raising awareness among the general public in order to improve the knowledge, attitudes, and behaviors of individual households and residents towards solid waste segregation practices. Furthermore, supportive policies, adequate infrastructure, and incentive mechanisms can also contribute to the wider adoption of these practices.

As a result of the literature review, it is evident that education, attitudes, behaviors, supportive policies, adequate infrastructure, and incentives are key factors that can enhance the source segregation situation in many countries, including the study area. The incentives that were found to have a significant impact included rebates on house taxes, water taxes, and fee-free solid waste collection, as these offerings led to a positive response from the respondents regarding the source segregation of waste in the study area.

6. Current scenario of waste management in Varanasi city

A2Z, a Gurugram, Haryana-based company, was given a 30-year contract in 2010 to manage solid waste in Varanasi City. However, the Varanasi Municipal Corporation (VMC) ended the business's contract because of anomalies. The Prime Minister's "*Swachchh Bharat Mission*" (Clean India Mission) in 2015 brought the city's solid waste management (SWM) initiative back to life, and the Central Government then signed an agreement with National Thermal Power Corporation (NTPC) to take up the SWM duty. Since 2016, M/s. IL & FS Environmental Infrastructure and Services Limited and Kiyana Solutions had been in charge of door-to-door garbage collection and transportation in 14 and 30 wards, respectively, but VMC ended its agreement with the companies in December 2019, and since then, the garbage was transported to the Karsara SWTP by the sanitary crew of the VMC; however, the waste management situation remained appalling.

A 7-year deal was signed by Varanasi Waste Solutions Private Limited (VWSPL), a company founded in 2020, which is a subsidiary of Antony Waste Handling Cell Limited (Thane-West, Maharashtra), with the VMC. Since its start on October 2, 2021, the MSWM project with VWSPL has been actively engaged in mechanized road sweeping, door-to-door MSW collection and transportation, and other associated services.

VWSPL was observed to be involved in solid waste management (SWM) in Varanasi City during a personal investigation. It oversees the implementation of door-to-door collection in almost ninety wards of the city. The use of small compartmentalized vehicles and the placement of large bins in public areas and residential colonies is in practice in the study area. In the morning, garbage tricycle rickshaws equipped with different color bins and auto tippers reach door-to-door and raise slogans to encourage the public to hand over their garbage to the waste collectors. They also appeal to the public not to mix wet and dry waste. However, the waste is collected in a mixed form by them and then dumped at transfer stations spread across the city

Figure 2.
Current waste management scenario in the study area.

before being sent to the *Karsada Waste Treatment Plant*, where it is further processed. The overall scenario is that the entire amount of refuse is collected and disposed of in mixed form, thereby defeating the ultimate goal of waste segregation. An example of waste management in the research area is shown in **Figure 2**.

7. Waste composition in Varanasi city

The composition and quantity of solid waste are influenced by several factors, including income, living standards, food preferences, local customs, climate, and resource availability [28]. A comparative analysis of the composition of MSW in Varanasi and other cities in India and around the world is provided in **Table 1**.

The waste characteristics of Varanasi have been compared to those of other Indian cities and towns. As shown in the table, the study area generates 0.428 kg of waste capita^{-1} day^{-1}, which is significantly less than that of other Indian and international cities. This can be attributed to the city's traditional nature, low quality of life, and early stages of economic growth, as there is a strong correlation between waste generation rates and income levels [39, 40].

Fifty households from all zones, as well as one transfer point from each zone (total eight), were chosen to calculate the average composition of Varanasi City's municipal solid waste (MSW). The net weight of each constituent was tallied, and the percentage value was determined using the following formula [32].

Cities/Countries	Organic/ Food	Paper/ Cardboard	Plastic	Metal	Glass	Textile	Disposable Nappies	Others	Generation Rate (kg capita^{-1} day^{-1})	Reference
Varanasi (India)	46.04	10.96	9.74	0.96	3.09	2.51	—	26.70	0.43	*Reported in the current study*
Delhi (India)	59.20	3.70	7.30	2.70	—	23.30	—	3.8	0.15–0.60	[29]
Kolkata (India)	50.56	6.07	4.88	0.19	0.34	—	—	37.96	0.58	[30]
Allahabad (India)	45.30	4.69	2.86	2.54	0.73	2.22	—	41.66	0.39	[31]
Puducherry (India)	38.40	30.00	10.40	4.10	5.00	4.50	—	7.60	0.59	[32]
Lagos (Nigeria)	56.00	14.00	4.00	4.00	3.00	—	—	19.00	0.63	[33]
Kitwe City (Zambia)	13.00	4.00	57.00	9.00	4.00	—	—	13.00	0.45	[34]
Kathmandu (Nepal)	67.80	6.50	0.30	4.90	1.30	19.20	—	—	0.497	[35]
London and Wales (UK)	20.20	33.20	10.20	5.7	9.30	2.10	3.9	8.10	—	[36]
Netherlands	37.00	22.00	14.00	4.04	4.30	3.60	6.60	8.00	—	[37]
USA	15.00	27.00	13.00	9.0	4.00	9.00	—	3.00	—	[38]

Source: Compiled by the author.

Table 1.
Evaluation of the MSW composition from different Indian and global cities.

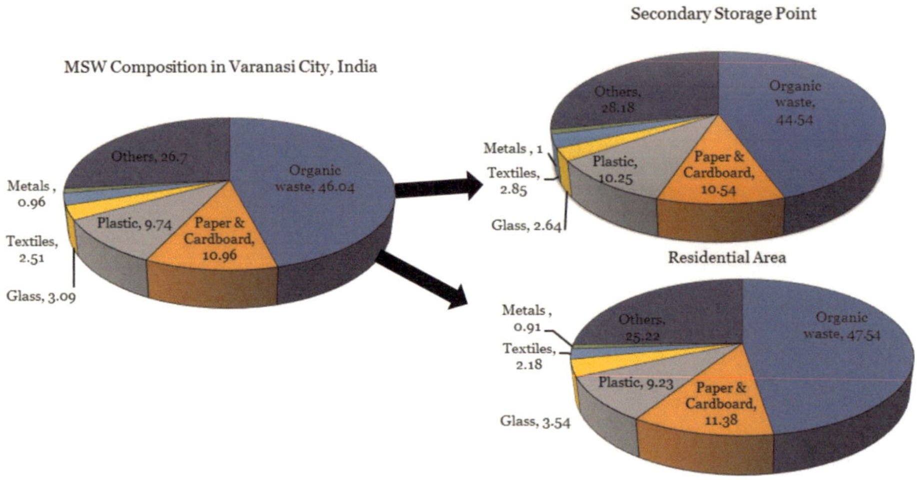

Figure 3.
MSW composition in Varanasi city.

Location	Organic waste	Paper & cardboard	Plastic	Glass	Textiles	Metals	Others
Varanasi City, India	46.04	10.96	9.74	3.09	2.51	0.96	26.70
Transfer Stations	44.54	10.54	10.25	2.64	2.85	1.00	28.18
Residential Area	47.54	11.38	9.23	3.54	2.18	0.91	25.22

Source: Field Survey, 2024.

Table 2.
Waste composition in Varanasi city.

$$\text{Waste in subcategory as a percentage} = \frac{\text{Net weight of the subcategory of waste}}{\textit{Total sample weight}} X\ 100$$

In terms of the physical composition of Varanasi's MSW, organic waste represents the largest percentage (46.04%), followed by paper (10.96%), plastic (9.74%), glass (3.09%), textile (2.51%), and metal (0.96%). The remaining 26.70% is made up of rubber, leather, bones, eggshells, soiled documents, dirt, earthen pots, and other inert materials (**Figure 3**).

The average value for Varanasi City was calculated using the values from residential areas and transfer stations. The result has been depicted in **Table 2**. The table indicates that Varanasi has a larger percentage of organic and food waste, which is common in developing countries. Despite being a developing country with a relatively low level of economic development, Varanasi City's waste production is positively correlated with economic growth.

It is also worth noting that, except for the UK and the Netherlands, most countries do not separate diapers, sanitary napkins, and other mixed biomedical waste for collection and disposal. In Varanasi, these items are also disposed of in the same manner. Since these materials are categorized as hazardous waste, it is imperative that they be stored separately and handled with utmost care and caution.

8. Results and discussion

The researcher developed a semi-structured interview schedule that was open only to residents of Varanasi City. Participants were asked to share any information and insights they had regarding the generation and segregation of garbage, as well as any

Characteristics	n = 440	%
Gender		
Male	266	60.45%
Female	174	39.55%
Age		
10–20	37	8.41%
21–30	209	47.50%
31–40	123	27.95%
41–50	55	12.50%
>50	16	3.64%
Educational qualification		
Doctorate	56	12.73%
Postgraduate (PG)	179	40.68%
Graduate	122	27.73%
Higher-secondary	23	5.23%
Matriculation	7	1.59%
Middle	20	4.55%
Professional	23	5.23%
Technical	7	1.59%
Other	3	0.68%
Occupation		
Agriculture	16	3.64%
Business person	35	7.95%
Govt. employee	54	12.27%
Private employee	73	16.59%
Professional	20	4.55%
Self-employment	39	8.86%
Student	161	36.59%
Unemployed	23	5.23%
Others	19	4.32%
Type of the family		
Joint	207	47.05%
Nuclear	233	52.95%
Monthly expenditures		

Characteristics	n = 440	%
< ₹ 5000	20	4.55%
₹ 5000–10,000	82	18.64%
< ₹ 10,000–15,000	81	18.41%
₹ 15,000–20,000	68	15.45%
> ₹ 20,000	189	42.95%

Source: Field Survey, 2024.

Table 3.
Demographic details of the respondents.

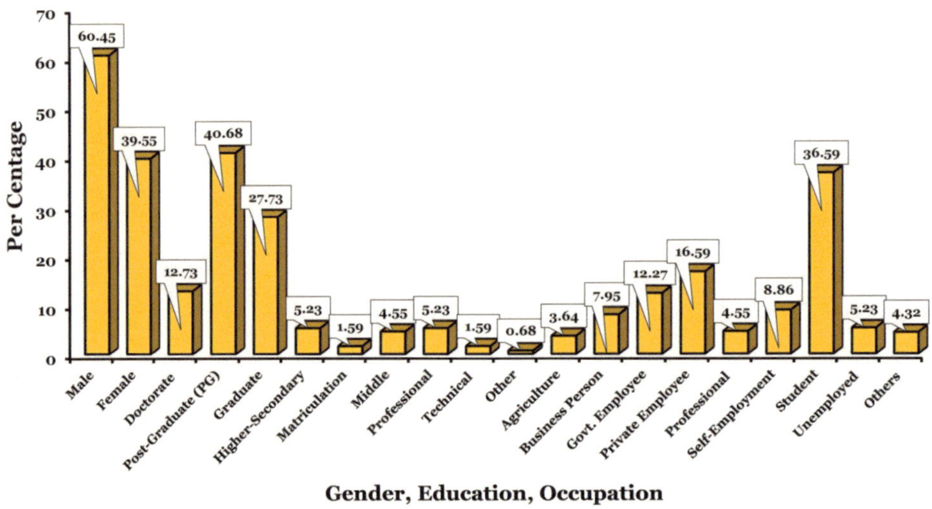

Figure 4.
Demographic details of the respondents.

obstacles that hindered these processes. Furthermore, they were requested to provide any details they had heard or observed about the waste management situation, with the aim of enhancing the understanding of solid waste management in the research area after the completion of the interview process.

The first segment of the schedule carries demographic details of the respondents, which have been portrayed in **Table 3** and **Figure 4**.

The demographic characteristics of the respondents are illustrated in **Table 1**, indicating that a total of 440 individuals participated in the interview process. Of these, 60.45% were male respondents, while 39.55% were female. This is likely due to male respondents feeling more comfortable discussing the topic with a male researcher. The majority of respondents, approximately 67%, were between the ages of 21 and 40, while only 3.64% were over the age of 50.

This suggests that younger individuals were more willing to participate and understood the challenges of source segregation and solid waste management. Moreover, the participants, comprising graduates, postgraduates, and students, exhibited a high level of engagement and enthusiasm in the survey. Another aspect of the demographic profile is that as expenditure (and, indirectly, income level) increases, individuals prefer to live in nuclear families, indicating that most families are residing in urban areas due to job constraints.

8.1 An overview of the source-based separation of solid waste

Based on the responses of the participants, it was evident that the majority (72.05%) were familiar with the concept of source segregation. However, simply familiarity with the term does not guarantee the effectiveness of the segregation programme. In a study conducted by Otitoju and Seng in Malaysia, the researchers also emphasized that mere acquaintance and information about source separation are insufficient to ensure the success of the agenda [41]. Abdel-Shafy and Mansour emphasized that community knowledge and involvement of diverse groups are crucial components for the success of any solid waste segregation strategy [38]. During the course of the interview process, certain respondents demonstrated an eagerness for significant transformation in the mindset of the individuals within the study area, aiming to affect a paradigm shift from the conventional use-and-discard mentality to the principles of the 3Rs (reduce, reuse, and refurbish). According to our discussions with the youth, they opined that practical application of knowledge is just truly effective when it is put into action at the grassroots level.

Three further questions were presented to the participants: their level of preparedness, the advantages they perceived, and any considerations they had for source separation. The response was overwhelmingly positive. Among the participants, 73.18% were contemplating it, 62.95% were aware of its benefit, and 84.09% were prepared to separate their waste materials at the source (**Figure 5**). So, where does the

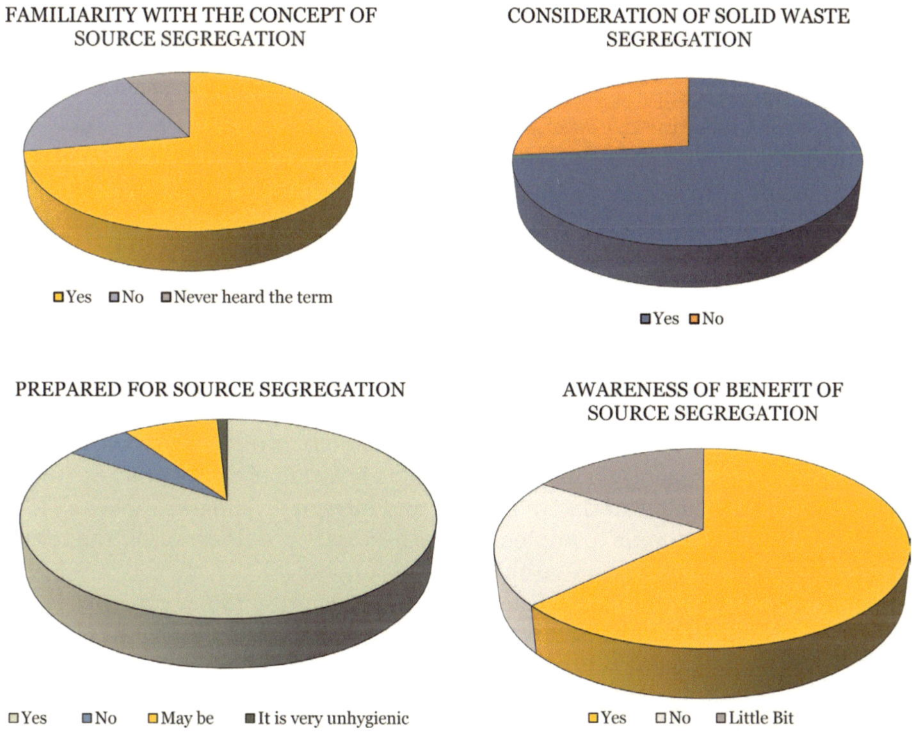

Figure 5.
Visual display of the responses.

failure lie? The researcher has endeavored to identify the underlying issues and obstacles that prevent source segregation from being successfully implemented.

In order to fulfill the aforementioned objective, three supplementary questions were posed to the respondents: whether there was a designated area for waste segregation in their residential community, whether they possessed separate containers for different types of solid waste at home, and whether there were separate containers in the outdoor area. Unfortunately, the responses did not demonstrate a high level of conviction. Specifically, 59.32% of the respondents reported the absence of a designated area for waste segregation in their residential community, while 48.18% indicated that they did not have separate containers at home. Additionally, 62.05% of the respondents indicated the absence of separate containers in the outdoor area. Further, during the nonparticipatory observation, it was noticed that separate containers with specific caution for dry or wet waste, compartmentalized vehicles, and transfer stations were loaded with mixed waste. As a result, the final outcome was futile, and all of the efforts were deemed ineffective (**Table 4**).

In addition to the previous question, the researcher asked to the respondents to indicate the reason for the lack of waste segregation in households. A total of 32.95% of the participants stated that the intermingling of waste by the municipal corporations made waste segregation at the household level unnecessary, as it was the primary cause of the failure of source separation (**Figure 6**). Meanwhile, 19.32% of the participants expressed that there was no facility to separate their waste; thus, they were unable to do so. Furthermore, 15.23% of the participants were not aware of the waste segregation process. As evident from the results, the majority of the respondents believed that waste segregation was not practical or useful for them. However, if they were provided with incentives, they would be willing to separate their waste.

The majority of respondents (84.55%) expressed a willingness to separate their waste if provided with economic incentives such as rebates on income or water taxes, fee-free door-to-door waste pickup, or other financial incentives. Research by Khanal et al. suggests that families who practice source segregation and deliver their waste separately may be eligible for reduced waste collection fees [42]. Alhassan et al. report that monetary incentives are the most effective motivator for source separation of solid waste [21]. Despite environmental concerns, Vining and Ebreo [43] found that people are more likely to be motivated by economic incentives to manage their recyclable items [43].

The respondents were given the opportunity to offer suggestions at the end of the interview to improve the MSWM program in the city. A significant number of respondents indicated that in order to bring about a positive change in the research area, it is imperative that the higher authorities, particularly the Mayor and the District Magistrate, display a strong sense of effective administrative leadership. Additionally, it is crucial that the staff members of the VMC who are responsible for waste management services are made aware of the significance of their work and discharge their duties with honesty and integrity. Gamaralalage et al. [5] also reported that it is simple to implement the necessary modifications if the relevant program has enough administrative and political support [5].

From conversing with the rag pickers, several noteworthy topics emerged. The majority of individuals who collect waste are Muslim and originally hail from West Bengal. They do not possess ration cards and are not included in the census due to their migratory lifestyle. With just their bare hands and feet, they commence their labor early in the morning, gathering abandoned and hazardous materials. Unfortunately, many rag pickers disregard recommended safety measures and fail to wear protective clothing, such as aprons, boots, masks, and gloves. As a result, they are at

Solid waste segregation practice by the survey respondents				
1	Queries	Responses	No. of Respondents	Percentage of Respondents
2	Are you familiar with the concept of segregating waste at the source of generation?	Yes	317	72.05%
		No	89	20.23%
		Never heard the term	34	7.73%
3	Have you ever considered segregating the solid waste that you generate?	Yes	322	73.18%
		No	118	26.82%
4	Are you aware of the benefits of segregating solid waste?	Yes	277	62.96%
		No	91	20.68%
		Little Bit	72	16.36%
5	Are you prepared to separate the solid waste in your home?	Yes	370	84.09%
		No	27	6.14%
		May be	39	8.86%
		It is very unhygienic	4	0.91%
6	Does your residential community have a designated area for waste segregation?	Yes	106	24.09%
		No	261	59.32%
		May be	31	7.05%
		I do not know.	42	9.55%
7	Do you have separate containers for different types of solid waste at your home?	Yes	198	45.00%
		No	212	48.18%
		Earlier, but not now	26	5.91%
		Other	4	0.91%
8	Do you have separate containers for different types of solid waste in your outdoor area?	Yes	139	31.59%
		No	273	62.05%
		Earlier, but not now	28	6.36%
		Other	—	
9	Do you sort the solid waste generated from your household before it exits your gate or door?	Yes	163	37.05%
		No	184	41.82%
		Some times	93	21.13%
10	Would you separate your waste if you receive incentives such as rebates on house or water taxes, or a fee-free door-to-door waste pickup service?	Yes	372	84.55%
		No	19	4.32%
		May be	42	9.55%
		Other	7	1.59%
11	What do you think is the reason for the lack of waste segregation in households?	No facilities to separate.	85	19.32%
		No need to separate, it is waste of time.	16	3.64%
		I am not aware, no information.	67	15.23%

Solid waste segregation practice by the survey respondents			
	Separation requires dust bins that can be bought at a high cost; thus, I cannot afford it.	19	4.32%
	My household produces a very small amount of waste; thus, separation is worthless.	85	19.32%
	Maintaining separate bins is problematic.	23	5.22%
	Municipal corporation mixes all waste, so what is the use of separating it.	145	32.95%
Source: Field Survey, 2024.			

Table 4.
Profile of waste segregation survey.

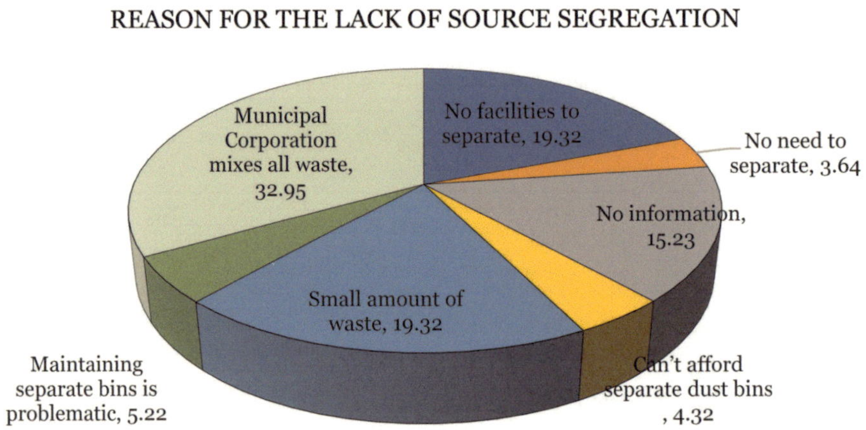

Figure 6.
Illustration of the difficulties with source separation.

risk of contracting diseases and infestations from direct contact with toxic hospital waste, batteries, carcasses, blood, pus, and sputum-soaked cotton. The smoke and putrid odor emanating from burning trash exacerbate the rag pickers' respiratory conditions. A small percentage of rag pickers also mentioned in their response that they are susceptible to dog bites.

Based on the information obtained from the interview process with key informants, it can be inferred that the interviewees have a good understanding of source separation, but they do not practice it, and if some of them do so, their waste is mixed in the collection and transportation process. This act demoralizes and deters them from separating the waste. The role of the rag pickers is also limited in the waste segregation process as they are not officially recognized in the system. The door-to-door collection of waste has improved the situation to some extent; however, to reach the ideal waste management scenario, a significant and arduous job needs to be done in the research area.

Extremely severe	Severe	Neutral	Less severe	Not severe at all
5	4	3	2	1
Challenges		Rank	Challenges	Rank
Lack of accountability and will power among staffs		5	Unavailability of data maintenance	4
Absence of a reuse, recycle, or take-back policy		5	Lack of awareness among residents	4
Lack of source segregation of waste		5	Insufficient support of the residents	4
Dearth of segregation of waste during handling		5	Lack of auditing of the work	4
Absence of cost evaluation		5	Lack of technical skill to manage waste	3
Rapid increase in waste generation		5	Insufficient infrastructure	1
Lack of training/human resource		4	Insufficient budget	1

Source: Field Survey, 2024.

Table 5.
Illustration of the challenges faced by VMC.

Furthermore, during focus group discussions with VMC staff, a list of challenges was offered along with a request to rate them on a five-point Likert scale. **Table 5** presents the information below.

The table reveals that VMC officials recognize that one of the major issues facing the city's waste management situation is the absence of source segregation and the failure to maintain waste segregation during collection and transportation stage. It is also vibrant that there is no issue with a lack of funding or inadequate infrastructure. Therefore, what is needed is strong administrative perseverance to better the state of affairs, as demonstrated in the Indian city of Indore, where Solid Waste Management By-Law, 2018 of the Indore Municipal Corporation has been put into effect by the city authorities to expedite source separation and 3R operations in the city [5].

9. Recommendations

Findings of the study indicate that the primary challenges are the absence of source segregation and the complete lack of waste segregation during collection and transportation. To improve waste segregation at the source while considering local factors, the following activities are recommended:

- Recognizing the critical role of research area residents as the primary waste generators, they must be incorporated from the project's outset. To achieve source separation of waste, promoting knowledge and cultivating a sense of civic responsibility are vital. To attain these objectives, collaborative efforts between local government agencies, NGOs, and other government bodies can be instrumental in launching extensive awareness campaigns.

- One of the most important tools in source segregation is awareness creation, which is the only mechanical equipment available for this purpose. Raising awareness can be facilitated by a variety of channels too, including print and electronic media, social media, and street plays.

- Additionally, the government must make it mandatory to incorporate waste education into curricula at all educational levels can help to ensure that people can come to know how to segregate waste to turn it into resource [44].

- Local authorities can adopt an informative and educational approach to promote source segregation. One effective method is to provide residents with information on the lifespan of different types of waste in their regional language. Additionally, implementing laws, utilizing catchphrases, disseminating animated videos on LED displays at various places in the city and on social media, offering cash incentives, and imposing penalties can all increase the likelihood of successful source segregation.

- Furthermore, to increase awareness, volunteers, home maids, housekeeping staff, and inhabitants of the research area can be trained to make a change [45]. If VMC adheres to the above specified meticulous protocols, favorable outcomes are possible.

- The promotion of the slogan "Har Din, Teen Bin (Every Day, Three Bins)" [45] in the research area is recommended to improve the segregation process at the domestic level. To this end, the VMC can develop and distribute specialized containers specifically intended for the collection of household waste, which cannot be used for any other purpose.

- Altering the beliefs and attitudes of individuals is a challenging task. Nonetheless, if the local authorities establish a target to disseminate the advantages of source segregation within each ward, it can be accomplished. To achieve this objective, the collaboration of NGOs and school children can be enlisted once again. It is essential to equip school children with various roles that will enable them to spread awareness among the residents of the research area effectively.

- The segregation process can be further enhanced by the design and distribution of such distinct containers intended only for household garbage collection.

- Altering the beliefs and attitudes of individuals is a challenging task. Nonetheless, if the local authorities establish a target to disseminate the advantages of source segregation within each ward, it can be accomplished. To achieve this objective, the collaboration of NGOs and school children can be enlisted once again. It is essential to equip school children with various roles that will enable them to disseminate awareness among the residents of the research area effectively (**Figure** 7).

- Waste separation in public spaces presents a significant challenge, but technological advancements offer assistance. To effectively separate trash, smart bins equipped with sensors and communication capabilities could be implemented in public areas. The exterior of the bin could be constructed with reinforced cement concrete (RCC), while the interior would feature a metal composition. The sensors would be attached to the RCC wall (**Figure 8**).

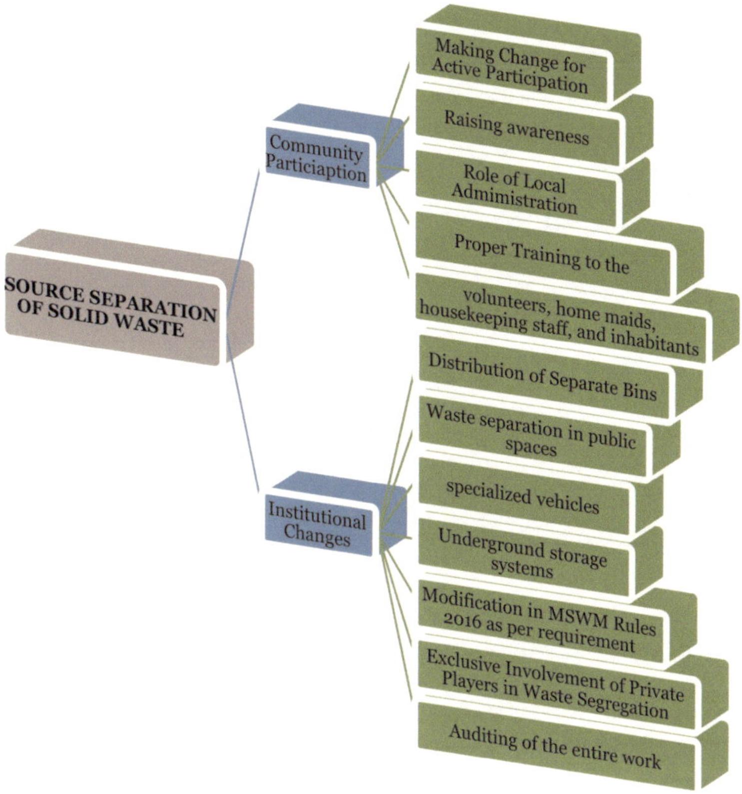

Figure 7.
Source separation success factors.

- According to VJ et al., inductive sensors can be used to identify metallic objects [46]. Similarly, glass, plastic, and fibers are separated using FT-IR spectroscopy, Hyper Spectral Imaging, NIR spectroscopy, multivariate analysis, and the X-ray fluorescence approach [47]. Rahman et al. propose that sensors based on near-infrared (NIR), mid-infrared (MID), and visual imaging sensor (VIS) technologies can be employed to identify paper waste [48].

- Additionally, a gas sensor and lid-mounted fill level indicator would ensure proper waste disposal by detecting the presence of gas and full bin status. Moreover, a central workstation connected to the municipal office could receive data from the microcontroller *via* Wi-Fi and sensors, as demonstrated by Dugdhe et al. [49].

- Incorporating solar panels onto rooftops can effectively address energy demands, while fencing around bins can maintain cleanliness in the surrounding area. Through the use of GIS network analysis, the workstation can recommend the most efficient routes for waste collection vehicles, thereby minimizing travel time and fuel consumption.

- Implementing these technological advancements can significantly enhance waste segregation and collection in public areas as depicted in **Figure 8**.

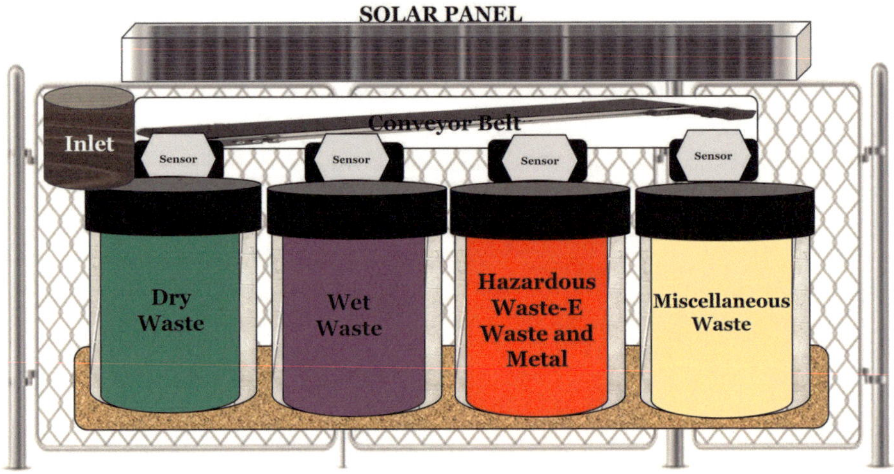

Figure 8.
Model for waste segregation bins in public areas.

- The collection of waste necessitates the presence of specialized vehicles, such as those designated for domestic, commercial, market, and hazardous waste, all of which must have assigned compartments. Upon collection, these vehicles carry the waste to transfer stations, where rag pickers and waste dealers can effectively sort the materials, with recyclables being sold at the end of the day. This method of waste disposal may significantly reduce the amount of waste sent to landfills, promoting recycling and reuse.

- To store wet or biodegradable waste, VMC can construct underground storage systems at transfer stations, as illustrated in **Figure 8**. These systems will prevent cattle, dogs, swine, and other animals from scattering litter, conceal the garbage, and eliminate foul odors, thereby enhancing the esthetics of the city.

- When establishing waste management regulations, it is crucial to consider local factors [5]. The MSWM Rules (2016) may be modified by the VMC in accordance with local needs in order to guarantee efficient waste management procedures (**Figure 9**).

- It is also essential to foster a collaborative and unified approach among the various departments responsible for waste management to prevent any potential miscommunication and to ensure the smooth operation of the entire waste management unit.

- Moreover, the VMC may consider employing waste management experts to design and implement source separation of waste materials, public bins, collection vehicles, and transportation methods tailored to the specific composition of waste generated in the area.

- Additionally, it is clear that thousands of individuals referred to as rag pickers in the study area rely on waste collection and recycling as a means of survival. Incorporating these individuals into SWM programs can be socially desirable, environmentally sustainable, and economically beneficial [50]. The VMC can

Figure 9.
Subterranean garbage storage model. Courtesy: HABA MEGA underground waste compactor.

easily include them officially at various housing societies, transfer stations, and treatment plants to segregate recyclables, which can then be recycled.

- An integral aspect of the waste management system must include thorough auditing encompassing all aspects from resource allocation to waste generation, collection, treatment, and final disposal. Higher authorities must enforce sanctions, such as reducing employee salaries or revoking revenue shares, against departments or individuals found responsible for lapses [51]. This ensures accountability and adherence to waste management protocols.

10. Conclusion

The separation of waste is a crucial aspect of an integrated and sustainable waste management system, as indicated by this study. The research identifies several significant obstacles that impede effective segregation of waste in the study area, including a lack of public awareness and the mixing of different types of waste without discrimination. Additionally, during the implementation of the study, various organizational and sociological challenges were encountered.

However, despite these challenges, the study offers practical solutions that are tailored to local conditions, which could prove beneficial to local governments. Waste management decision-makers should consider planning or refining source separation initiatives with a strong emphasis on policy development and understanding the potential impacts of these programs. Incorporating evidence-based insights from studies such as this one can enhance the effectiveness of these initiatives. Overall, by addressing these challenges and implementing informed strategies, municipalities can improve their waste management practices and promote sustainable outcomes within their communities.

Acknowledgements

Under the Post-Doctoral Fellowship for the Year 2022–2023 with the F. No. 3-138/2022-2023/PDF/GEN, the Indian Council of Social Science Research (ICSSR), New Delhi, has provided funding for this study.

Author details

Awadhesh Kumar* and Narendra Kumar Rana
Department of Geography, Institute of Science, Banaras Hindu University, Varanasi, India

*Address all correspondence to: awkumar14@gmail.com

References

[1] Ma J, Hipel KW, Hanson ML, Cai X, Liu Y. An analysis of influencing factors on municipal solid waste source-separated collection behavior in Guilin, China by using the theory of planned behavior. Sustainable Cities and Society. 2018;**37**:336-343. DOI: 10.1016/j.scs.2017.11.037

[2] Chang NB, Pires A. Sustainable Solid Waste Management: A Systems Engineering Approach. The institute of Electrical and Electronics Engineers, Inc. John Wiley & Sons; 2015

[3] Sharma KD, Jain S. Municipal solid waste generation, composition, and management: The global scenario. Social Responsibility Journal. 2020;**16**(6): 917-948

[4] Rana R, Ganguly R, Gupta AK. An assessment of solid waste management system in Chandigarh City, India. Electronic Journal of Geotechnical Engineering. 2015;**20**(6):1547-1572

[5] Dickella Gamaralalage PJ, Ghosh SK, Onogawa K. Source separation in municipal solid waste management: Practical means to its success in Asian cities. Waste Management & Research. 2022;**40**(3):360-370. DOI: 10.1177/0734242X211049

[6] Chen H, Yang Y, Jiang W, Song M, Wang Y, Xiang T. Source separation of municipal solid waste: The effects of different separation methods and citizens' inclination—Case study of Changsha, China. Journal of the Air & Waste Management Association. 2017;**67**(2):182-195. DOI: 10.1080/10962247.2016.1222317

[7] Oduro-Kwarteng S, Anarfi KP, Essandoh HM. Source separation and recycling potential of municipal solid waste in Ghana. Management of Environmental Quality: An International Journal. 2016;**27**(2):210-226. DOI: 10.1108/MEQ-03-2015-0038

[8] Yang L, Li ZS, Fu HZ. Model of municipal solid waste source separation activity: A case study of Beijing. Journal of the Air & Waste Management Association. 2011;**61**(2):157-163. DOI: 10.3155/1047-3289.61.2.157

[9] Ministry of Environment FA. Solid Waste Management Rules, 2016. New Delhi; 2016

[10] Rana Kaseva ME, Mbuligwe SE, Kassenga G. Recycling inorganic domestic solid wastes: Results from a pilot study in Dar es Salaam City, Tanzania. Resources, Conservation and Recycling. 2002;**35**(4):243-257

[11] Bennagen EC, Nepomuceno G, Ramil C. Solid Waste Segregation and Recycling in Metro Manila: Household Attitudes and Behavior. Singapore, SG: EEPSEA; 2002

[12] Yadav V, Karmakar S. Sustainable collection and transportation of municipal solid waste in urban centers. Sustainable Cities and Society. 2020;**53**: 101937. DOI: 10.1016/j.scs.2019.101937

[13] Singh, Rana PB, Rana PS. The Ganga Ghats, Varanasi: The Riverfront Landscape. Varanasi: Banaras (Varanasi) Tara Book Agency; 1999

[14] Kumar EB, Chaturvedula S, Thakur R, Singh R, Sekhar RA. Holistic Waste Management Strategy for Varanasi City. Varanasi: UN Environmental Programme; 2018

[15] Chakrabarty R. GIS for Effective Waste Management in City of Varanasi

[Internet]. 2024. Available from: https://www.esri.in/en-in/newsroom/blog/gis-for-effective-waste-management-in-city-of-varanasi/.

[16] Tiwari SK Final Report of Ward Delimitation of Varanasi Municipal Corporation. Varanasi: Dainik Jagran; 2022. Available from: https://www.jagran.com/uttar-pradesh/varanasi-city-final-report-of-ward-delimitation-of-varanasi-municipal-corporation-released-26-wards-merged-and-30-abolished-23195791.html

[17] Bartlett James EII, Kotrlik JW, Higgins CC. Organizational research: Determining appropriate sample size in survey research appropriate sample size in survey research. Information Technology, Learning, and Performance Journal. 2001;**19**(1):43

[18] Israel GD. Determining Sample Size. A series of the Program Evaluation and Organizational Development. University of Florida; 1992

[19] Babazadeh T, Nadrian H, Mosaferi M, Allahverdipour H. Identifying challenges and barriers to participating in the source separation of waste program in Tabriz, Northwest of Iran: A qualitative study from the citizens' perspective. Resources. 2018;7(3):53

[20] Babazadeh T, Nadrian H, Mosaferi M, Allahverdipour H. Challenges in household solid waste separation plan (HSWSP) at source: A qualitative study in Iran. Environment, Development and Sustainability. 2020; **22**:915-930. DOI: 10.1007/s10668-018-0225-9

[21] Alhassan H, Kwakwa PA, Owusu-Sekyere E. Households' source separation behaviour and solid waste disposal options in Ghana's Millennium City. Journal of Environmental Management. 2020;**259**:110055. DOI: 10.1016/j.jenvman.2019.110055

[22] Oduro-Appiah K, Aggrey BE. Determinants of Source Separation of Municipal Solid Waste in Developing Countries: The Case of Ghana. 2013

[23] Zhang W, Che Y, Yang K, Ren X, Tai J. Public opinion about the source separation of municipal solid waste in Shanghai, China. Waste Management & Research. 2012;**30**(12):1261-1271. DOI: 10.1177/0734242X12462277

[24] Singh A, Raj P. Segregation of waste at source reduces the environmental hazards of municipal solid waste in Patna, India. Archives of Environmental Protection. 2018;**44**(4):96-110. DOI: 10.24425/122306

[25] Rathore P, Sarma SP. Investigation of factors influencing source separation intention towards municipal solid waste among urban residents of India. Resources, Conservation and Recycling. 2021;**164**:105164. DOI: 10.1016/j.resconrec.2020.105164

[26] Matsumoto S. Waste separation at home: Are Japanese municipal curbside recycling policies efficient? Resources, Conservation and Recycling. 2011;**55**(3):325-334. DOI: 10.1016/j.resconrec.2010.10.005

[27] Adefris W, Damene S, Satyal P. Household practices and determinants of solid waste segregation in Addis Ababa city, Ethiopia. Humanities and Social Sciences Communications. 2023;**10**(1):1-10. DOI: 10.1057/s41599-023-01982-7

[28] Jin J, Wang Z, Ran S. Solid waste management in Macao: Practices and challenges. Waste Management. 2006;**26**(9):1045-1051

[29] Mor S, Ravindra K, De Visscher A, Dahiya RP, Chandra A. Municipal solid waste characterization and its assessment for potential methane generation: A case study. Science of the Total Environment. 2006;**371**(1–3):1-10

[30] Chattopadhyay S, Dutta A, Ray S. Municipal solid waste management in Kolkata, India–A review. Waste Management. 2009;**29**(4):1449-1458

[31] Sharholy M, Ahmad K, Vaishya RC, Gupta RD. Municipal solid waste characteristics and management in Allahabad, India. Waste Management. 2007;**27**(4):490-496

[32] Pattnaik S, Reddy MV. Assessment of municipal solid waste management in Puducherry (Pondicherry), India. Resources, Conservation and Recycling. 2010;**54**(8):512-520

[33] Ogwueleka T. Municipal solid waste characteristics and management in Nigeria. Journal of Environmental Health Science & Engineering. 2009; **6**(3):173-180

[34] Mwanza BG, Mbohwa C, Telukdarie A. Municipal solid waste management in Kitwe City: An engineering management perspective. Management of Environmental Quality: An International Journal. 2018;**29**(6): 1075-1092

[35] Dangi MB, Pretz CR, Urynowicz MA, Gerow KG, Reddy JM. Municipal solid waste generation in Kathmandu, Nepal. Journal of Environmental Management. 2011; **92**(1):240-249

[36] Burnley SJ. A review of municipal solid waste composition in the United Kingdom. Waste Management. 2007; **27**(10):1274-1285

[37] Goorhuis M, Reus P, Nieuwenhuis E, Spanbroek N, Sol M, van Rijn J. New developments in waste management in the Netherlands. Waste Management & Research. 2012;**30**(9_suppl):67-77

[38] Abdel-Shafy HI, Mansour MS. Solid waste issue: Sources, composition, disposal, recycling, and valorization. Egyptian Journal of Petroleum. 2018; **27**(4):1275-1290

[39] Kumar KN, Goel S. Characterization of municipal solid waste (MSW) and a proposed management plan for Kharagpur, West Bengal, India. Resources, Conservation and Recycling. 2009;**53**(3):166-174

[40] Srivastava V, Ismail SA, Singh P, Singh RP. Urban solid waste management in the developing world with emphasis on India: Challenges and opportunities. Reviews in Environmental Science and Bio/ Technology. 2015;**14**:317-337

[41] Otitoju TA, Seng L. Municipal solid waste management: Household waste segregation in Kuching South City, Sarawak, Malaysia. American Journal of Engineering Research (AJER). 2014; **3**(6):82-91

[42] Khanal A, Giri S, Mainali P. The practices of at-source segregation of household solid waste by the youths in Nepal. Journal of Environmental and Public Health. 2023;**2023**(1):5044295. DOI: 10.1155/2023/5044295

[43] Vining J, Ebreo A. What makes a recycler? A comparison of recyclers and nonrecyclers. Environment and Behavior. 1990;**22**(1):55-73. DOI: 10.1177/0013916590221003

[44] Kumar A. Dumping yard-free society through sustainable solid waste

management: A case study of Varanasi City. In: Waste Management and Resource Efficiency: Proceedings of 6th IconSWM 2016. Singapore: Springer; 2019. pp. 409-421

[45] Alag Karo Program: An innovation-driven approach to waste segregation in Gurugram" by Smita Singh, March 30, 2020, https://csrbox.org/Impact-stories_full_Alag-Karo-Program–An-innovation-driven-approach-to-waste-segregation-in-Gurugram_206

[46] Aleena VJ, Balakrishnan K, Rosmi TB, Krishna KS, Sreejith S, Subha TD. Automatic waste segregator and monitoring system. Journal of Microcontroller Engineering and Applications. 2016;**3**(2):1-7

[47] Sujatha K, Radha RC. Technologies for segregation and management of solid waste: A review. In: 2016 International Conference on Emerging Trends in Engineering, Technology and Science (ICETETS). Pudukkottai, India: IEEE; 2016. pp. 1-4

[48] Rahman MO, Hussain A, Basri H. A critical review on waste paper sorting techniques. International Journal of Environmental Science and Technology. 2014;**11**:551-564

[49] Dugdhe S, Shelar P, Jire S, Apte A. Efficient waste collection system. In: 2016 International Conference on Internet of Things and Applications (IOTA). Pune, India: IEEE; 2016. pp. 143-147

[50] Oteng-Ababio M. The role of the informal sector in solid waste management in the GAMA, Ghana: Challenges and opportunities. Tijdschrift voor Economische en Sociale Geografie. 2012;**103**(4):412-425. DOI: 10.1111/j.1467-9663.2011.00690.x

[51] Swaminathan M. How can India's waste problem see a systemic change. Economic and Political Weekly. 2018;**53**(16):1-20

Chapter 10

Strategies for Enhancing the Efficiency of Packaging and Managing Packaging Waste in Compliance with Regulations

Enrico Maria Mosconi, Andrea Colantoni and Mariarita Tarantino

Abstract

The packaging industry is a critical key point of the global economy, estimated to be worth around $1 trillion dollars. It is made of different sectors that work together to create a dynamic value chain, connecting companies, manufacturers, finance, and environmental aspects that influence each other. Overall, from the upstream to downstream of the processes, the packaging industry is a significant indicator of the economies' development, production, and consumption trends. It is expected to grow by 4% annually from 2021 to 2026, mainly due to the increased use of materials such as cardboard. But the underrated aspect during the industrialization period has become a priority. The addressing of the environmental issues is driving a growing shift toward environmentally friendly packaging materials due to increasing regulatory pressure today. It is essential to adopt a circular approach in developing new products that considers both the economic and eco-design aspects of the evolving packaging industry. New industrial trends strongly complement the environment with the transition of the industry. The purpose of this chapter is to address the dynamics of the packaging industry in light of the environmental transition and the industrial circular economy approach.

Keywords: packaging, waste, circular economy, European regulation, industry, recyclability, reusability

1. Introduction

1.1 The packaging and packaging industry for a circular economy

In the current context, sustainability and sustainable development are central issues in global policies, and attention is increasingly shifting toward environmentally friendly practices. Sustainability implies a holistic approach that considers the interaction of the three fundamental pillars, namely, environmental, social, and economic, with the aim of satisfying the needs of the present generation without compromising those of the future generation. In this scenario, packaging plays a fundamental role. Originally designed only to protect and transport products, today's packaging design and manufacturing must take environmental impact into account. Packaging is an essential component in many industries, including supermarkets, restaurants, and pharmacies. This sector has seen significant growth globally, playing a dynamic role in the world economy and with a rapid increase in its market value. The scientific community's definition of packaging has evolved over time. It now refers to any material that is used to contain, protect, deliver, or preserve products. The primary purpose of packaging is to safeguard products during distribution, storage, transport, trade, and use [1]. Additionally, packaging should be designed to allow for reuse whenever possible [2].

1.1.1 The role of the packaging industry in the global economy

The increase in packaging consumption is influenced by improved well-being and changes in people's lifestyles, particularly in food consumption. Packaging consumption affects not only the final consumer but also all sectors involved in the production of goods and services. The packaging industry, valued at about $1 trillion, encompasses manufacturers, companies, finance, and environmental aspects, creating a complex interdependent ecosystem. Its significance continues to grow and significantly contributes to global economic development, production, and consumption trends. The industry is expected to experience an annual growth rate of 4% in 2026, largely due to the increasing demand for packaging materials, such as cardboard, plastic, metal, glass, and wood. The food and beverage, pharmaceutical, and cosmetic industries are the main sectors utilizing packaging materials, with the food and beverage industry accounting for 50–60% of the total market. In 2023, the global packaging market was valued at around $1.2 trillion and is projected to reach approximately $1.6 trillion by 2032, with a compound annual growth rate (CAGR) of 4.2% (**Figure 1**).

In Europe, 27.7% of total packaging and 44% of plastic packaging are produced in the agri-food sector, representing 20–25% of food production costs. The European packaging market was valued at €200 billion in 2023, with significant portions of the market occupied by paper and cardboard (40%), followed by plastic (35%), glass (15%), metal (10%), and other materials (5%). **Table 1** summarizes the market share

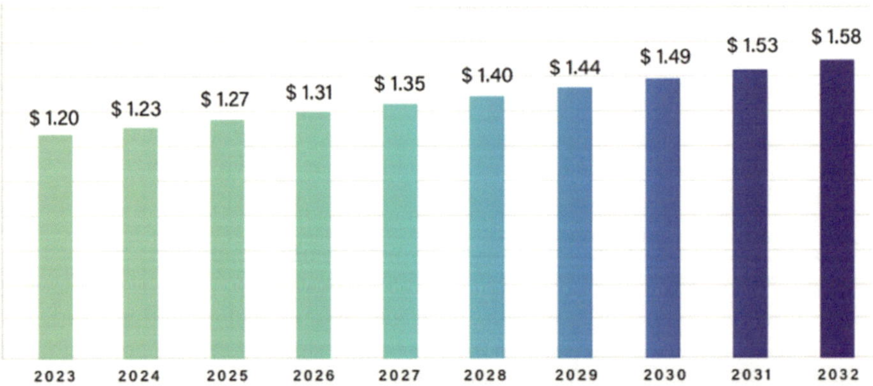

Figure 1.
Packaging market size to 2032 (USD in trillion). Source: Towards Packaging.

Region	Market value	Plastic	Paper and cardboard	Metal	Glass	Other materials
Global	$1.2 trillion	$388 billion (37%)	$357 billion (34%)	$158 billion (15%)	$110 billion (10%)	$42 billion (4%)
Europe	€200 billion	€70 billion (35%)	€80 billion (40%)	€20 billion (10%)	€30 billion (15%)	€10 billion (5%)
Italy	€30 billion	€10.5 billion (35%)	€12 billion (40%)	€3 billion (10%)	€4.5 billion (15%)	€1.5 billion (5%)

Table 1.
Market shares in the packaging industry by region and material in 2023. Sources: Processing based on data European Commission, Towards Packaging.

of the global, European, and Italian packaging markets in 2023, broken down by material types.

With technological advancements, global supply chains have grown longer, leading to increased packaging use and waste generation [3]. Packaging waste constitutes a significant portion, ranging from 15% to 20%, of total municipal solid waste in many countries [4].

As per the future estimates by the European Commission, the production of packaging waste is expected to rise by 107 million tons by 2040, which will have significant impacts on both the environment and the economy, including the well-known effects of greenhouse gases [5]. Although packaging is an essential component of the supply chain and economy, it lacks proper coordination in terms of design, use, and end-of-life decisions. The agri-food sector is particularly affected by packaging problems. In Europe, 27.7% of total packaging and 44% of total plastic packaging are produced in this sector. In addition, packaging accounts for between 20% and 25% of food production costs [6]. The recovery rate of packaging waste in Italy has seen a significant increase over the years. In 1998, it was 33.2%, which rose to 68.6% by 2008 and further increased to 74% by 2012 [7]. Recently, in 2022, the recycling rate reached 71.5%, and the total recovery rate reached 80.5% [8]. This demonstrates a continuous commitment and progress in the sector.

1.1.2 Challenges and opportunities in transitioning to a circular economy

To minimize the problem and optimize this crucial industry, there is a need for better design, usage, and end-of-life decisions [9]. Sustainable packaging materials are those that reduce the use of virgin resources and are made from readily available recyclable or reusable postconsumer materials [10]. The sustainability of packaging materials depends on a range of factors, including economic and environmental costs and impacts, the functionality of esthetic properties, and the effects of production and end-of-life processing on both local and global scales [11]. Unsustainable packaging materials have harmful effects on both the environment and the shelf life of the product. Therefore, the European Commission has taken a significant step by defining a comprehensive regulation, poised to replace the previous Directive on Packaging (2018/852/EU), enabling more coherent oversight among European member states. This regulation sets forth requirements spanning the entire lifecycle of packaging, from material sourcing to ultimate disposal, facilitating its market introduction. The Packaging and Packaging Waste Regulation aims to address some major concerns such as the excessive use of packaging, which is also known as overpackaging; the surge in the consumption of disposable products; and the challenges faced by internal markets due to unclear and noncompliant labeling. The European Commission has taken a range of actions to tackle the challenges of environmental sustainability and resource management beyond just recycling. They have set reduction targets that involve redesigning products, promoting reuse and refill, and implementing specific requirements for reuse and recycling in certain sectors. It is essential to consider the number of times a product can be reused and recycled rates [12]. With approximately one-third of packaging not being recyclable, it is crucial to introduce clear guidelines. This approach demonstrates a significant commitment to achieving a circular economy [13]. The legislative agenda extends beyond mere circularity, striving to not only protect the environment but also invigorate the European single market in green sectors. The reduction of material usage and promotion of recycling, reuse, and innovation, such as compostable packaging and eco-design principles, are the strategies to achieve the objectives [14]. The effectiveness of these goals is partly realized through the application of regulations influencing extended producer responsibility (EPR). EPR covers the entire supply chain, including production, logistics, and distribution. The scientific literature demonstrated that only recently have manufacturers started to address the overpackaging issue, adopting concrete initiatives aimed at designing more environmentally sustainable outer packaging [15]. Given scientific and technological advancements, the integration of circularity principles into packaging design and production processes, supported by legal instruments such as regulations, also compels the industry to implement policies promoting the adoption of eco-design and sustainable production practices, representing a fundamental step toward more efficient and sustainable resource management [16]. In addition, environmental pressures, such as the use of materials and the generation of waste associated with disposable packaging, require a shift away from single-use packaging.

1.2 The logistics of the packaging world

The world of packaging logistics is increasingly guided by the principles of the circular economy and sustainable development. It involves managing packaging materials throughout the supply chain, from production to end-of-life disposal or recycling. This includes transportation, storage, and handling of packaging materials, as well as the processes involved in returning used packaging for reuse or recycling. Globalization has extended supply chains, leading to increased packaging use and substantial waste generation. Packaging waste makes up a significant portion of municipal solid waste, ranging from 15% to 20% in many countries. Reverse logistics, which involves the return and recycling of used packaging, plays a crucial role in creating environmental, economic, and social benefits. This supports the circular economy by minimizing resource wastage through the reuse and recycling of packaging materials.

1.2.1 Addressing challenges and driving innovations in packaging logistics

Packaging logistics also need to address regulatory challenges and environmental impacts. For example, the European Union (EU) has set targets for reducing packaging waste and improving recycling rates. By 2030, all packaging on the EU market must be reusable or economically recyclable. This requires significant investment in recycling infrastructure and technological innovations to handle diverse materials effectively. In addition to economic and regulatory factors, packaging logistics are also influenced by cultural and social considerations. Consumer behavior and societal attitudes toward recycling and sustainability significantly impact the effectiveness of packaging logistics. Educational campaigns and clear labeling can enhance consumer participation in recycling programs. To improve the efficiency of packaging logistics, stakeholders must adopt holistic approaches that integrate environmental, social, and economic objectives. This includes investing in advanced technologies for recycling, redesigning packaging to reduce waste, and fostering collaborative efforts among governments, businesses, and consumers to promote sustainable practices. Managing the complex interdependencies between packaging and logistics systems requires these practices to be crucial.

Packaging is essential for ensuring the efficiency and functionality of production, logistics, and marketing systems in the industrial sectors. It plays a critical role in transforming raw materials into finished products that can be easily distributed globally. Packaging solutions are vital for every physical goods industry as they protect, store, and transport products. The packaging industry is closely linked to the logistics industry. Additionally, the packaging industry is intricately connected to environmental performance. Environmentally friendly packaging plays a crucial role in optimizing space in warehouses and transport vehicles, thereby reducing logistics costs, cutting down on CO_2 emissions, and enhancing supply chain efficiency in logistics. It is important to note that the logistics sector requires products that are easily transportable and efficiently stored. In a circular economy, a reverse logistics system is necessary to handle the recycling and remanufacturing processes until new products are created.

The management of product development throughout the supply chain to maximize the remaining value after product disposal is known as reverse logistics. In 2024, the global logistics packaging market is estimated to be valued at $24.5 billion. Looking ahead, the market has the potential to reach $33.2 billion by 2034, expanding at a CAGR of 3.1% (**Table 2**).

This growth highlights the growing importance of sustainable and innovative packaging solutions in the global supply chain. This is driven by technological advancements and regulatory pressures to minimize environmental impact.

The efficiency of logistical systems is significantly impacted by packaging. However, many packaging designers often overlook the costs associated with packaging in the logistical system. Packaging specifications directly affect the time needed for packaging operations, which in turn impacts product lead time and delivery performance to the customer. The relationship between packaging and logistical activities is illustrated in **Table 3**.

Attributes	Key insights
Estimated logistics packaging market size (2024)	$ 24.5 billion
Projected logistics packaging market valuation (2034)	$ 33.2 billion
Value-based CAGR (2024–2034)	3.1%

Table 2.
Estimation of logistics packaging market potential 2024–2034.

Logistics activity	Trade-offs
Transportation	
Increased package information	Decreases shipment delays; increases package information; decreases tracking of lost shipments.
Increased package protection	Decreases damage and theft in transit but increases package weight and transport costs.
Increased standardization	Decreases handling costs and vehicle waiting time for loading and unloading; increases standardization; increases modal choices for the shipper and decreases the need for specialized transport equipment.
Inventory	
Increased product protection	Decreases theft, damage, and insurance; increases product availability (sales); increases product value and carrying costs.
Warehousing	
Increased package information	Decreases order filling time and labor cost.
Increased product protection	Increases cube utilization (stacking) but decreases cube utilization by increasing the size of the product dimensions.
Increased standardization	Decreases material handling equipment costs.
Communications	
Increased package information	Decreases other communications about the product.

Table 3.
Packaging cost trade-offs with other logistics activities.

1.3 Standards for the packaging industry

ISO standards are internationally recognized specifications for products, services, and systems, ensuring quality, safety, and efficiency, which are crucial to the packaging industry. Key ISO standards are important in the packaging industry, along with specific standards tailored to packaging within certain industries that have stricter regulations. ISO 9001 is a crucial quality management standard for many businesses. For packaging makers, it demonstrates the ability to produce packaging of ISO quality, indicating that the company operates a comprehensive quality management system. The management team can utilize the eight quality management principles of the ISO 9001 standard to enhance performance throughout the business. ISO 14001 certification highlights a company's dedication to environmental sustainability. Environmental impact is a major concern for many packaging companies, and the ISO 14001 accreditation promotes best practices by encouraging companies to actively reduce waste through efficient processes, design packaging with minimal material usage, and use recycled and recyclable materials whenever possible. There are also more specific ISO standards related to packaging in various industries. For example, ISO 15378 focuses on primary packaging materials for medicinal products, outlining specific requirements for applying ISO 9001 concerning good manufacturing practices. These standards aid medicinal packaging firms in minimizing safety risks and product contamination and ensuring product efficacy and durability. ISO 8317 pertains to companies producing child-resistant packaging and includes mechanical testing methods for resealable packaging systems. ISO 16106 deals with transport packaging for hazardous materials, incorporating plastic compatibility tests for packaging and intermediate bulk containers. The ISO 18600 to ISO 18604 standards offer comprehensive guidelines for environmentally friendly packaging, covering various aspects of the packaging lifecycle. ISO 18600 provides general requirements and a framework for assessing the environmental impact of packaging. ISO 18601 focuses on optimizing the packaging system to reduce environmental impact. ISO 18602 addresses the reuse of packaging, while ISO 18603 pertains to material recycling. ISO 18605 focuses on energy recovery, and ISO 18604 covers organic recycling. These standards help organizations design packaging that is resource-efficient and environmentally sustainable throughout its lifecycle.

The International Safe Transit Association (ISTA) has developed test protocols to simulate the environmental hazards encountered during the shipment and handling of packaged products. These protocols are designed to ensure that the packaging can withstand the rigors of transit and protect its contents. The key protocols include:

ISTA 1A: Non-simulation integrity performance test for packaged products weighing 150 pounds (68 kg) or less.

ISTA 3A: General simulation performance test for packaged products weighing 150 lb. (68 kg) or less.

ISTA 6-Amazon.com-SIOC: Simulation test for products shipped in their containers, tailored to the Amazon implementation network.

By following these protocols, product damage and loss during transport can be reduced, ensuring the integrity of the product upon arrival.

GS1 standards improve supply chain efficiency and visibility with unique identification systems such as barcodes and radio frequency identification (RFID) tags. Key components include the Global Trade Item Number (GTIN) for product identification, Serial Shipping Container Code (SSCC) for tracking logistics units, Global Location Number (GLN) for identifying locations, and Global Data Synchronization Network (GDSN) for sharing standardized product information. The International Plant Protection Convention (IPPC) and International Standards for Phytosanitary Measures (ISPM), particularly ISPM 15, aim to prevent the international spread of pests through wood packaging materials. Key measures include heat treatment and methyl bromide fumigation to ensure that wood packaging materials used in international trade are treated to prevent the spread of invasive species. ASTM standards for packaging, developed by the American Society for Testing and Materials, cover various aspects of packaging materials and performance, such as performance testing of shipping containers and systems, drop tests of loaded containers by free fall, and determining compressive resistance of shipping containers, components, and unit loads. European Organisation for Technical Approvals of Packaging and Packaging Systems (EUMOS) standards, set by the European Safe Logistics Association, focus on ensuring the safety and stability of cargo during transport. Key standards include certification of cargo securing systems and testing of cargo securing devices. By following these standards, packaging materials and processes not only effectively protect products but also promote sustainability and safety throughout the supply chain. This approach ultimately enhances logistical efficiency, reduces environmental impact, and fosters consumer trust.

2. Issues and problems of packaging

The challenge of the circular economy is to minimize material waste by keeping materials in use for as long as possible. This involves closing material loops to retain the qualities needed for the original function. Various approaches, such as the nine

"R's" [17], focus on retaining the highest economic value by reusing products and materials. By extending the lifecycle of products and materials, the need for new materials is reduced, which in turn lessens the environmental impact of material use. Sustainable practices, such as minimizing energy use to close the loops, are crucial for achieving this (**Figure 2**).

2.1 High use of virgin materials and waste growth

Packaging plays a critical role in safely distributing products in today's society and supply chains. In Europe, it is a major consumer of new materials. Conventional packaging relies on extracting virgin resources like plastics, paper, metals, and glass from the natural environment during production. After products reach consumers, most of the packaging is disposed of in landfills or becomes litter. Additionally, emissions are released during transportation between manufacturing facilities, filling lines, and warehouses. With global packaging demand estimated to increase by 3–4% annually, the environmental impact of packaging is expected to grow.

In 2022, the packaging sector was responsible for 40% of total plastic use [18] and 50% of paper use in the EU [19]. The reliance on virgin materials for packaging significantly contributes to resource extraction and environmental degradation. For instance, the production of virgin plastic is energy-intensive and leads to greenhouse gas emissions. The extraction and processing of raw materials for packaging, including metals and glass, also have significant environmental impacts, such as habitat destruction, soil and water pollution, and high energy consumption [20]. Packaging waste is a major issue in many countries, constituting 30–35% of the municipal solid waste produced annually in industrialized nations. Despite the establishment of specific recycling and recovery targets, this type of waste has continued to grow over the years, especially in Europe [21]. In 2021, the European Union produced an average of 188.7 kg

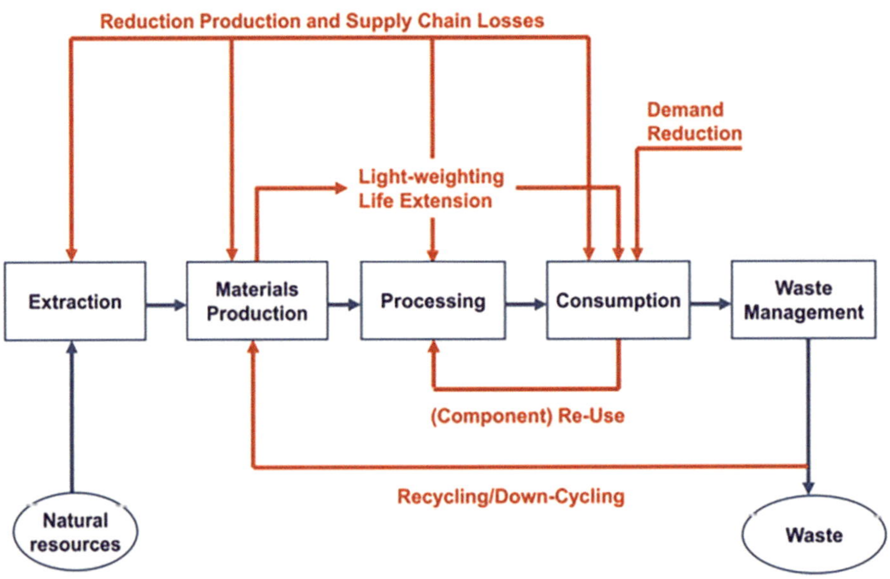

Figure 2.
Opportunities for material efficiency. Source: Sustainability of reusable packaging—Current situation and trends.

of packaging waste per person, with significant variations among countries [22]. The rise of online shopping and convenience-focused consumption patterns has worsened the situation, resulting in a surge in single-use packaging.

The recycling rates for certain materials, such as paper and metals, are relatively high; however, the recycling rate for plastics in the EU remains below 50% [23]. This is due to several challenges associated with recycling plastic packaging, including the diversity of plastic types, contamination issues, and the economic viability of recycling processes. Furthermore, the infrastructure for collecting, sorting, and processing recyclable materials is often inadequate, resulting in high rates of incineration and landfill disposal.

The environmental impact of packaging extends beyond waste generation. The production and disposal of packaging materials contribute to pollution and resource depletion. Making plastic packaging generates a significant amount of CO_2 emissions, contributing to climate change [24]. Improper disposal of packaging waste, particularly plastics, leads to ocean pollution, resulting in damage to marine life and ecosystems [25]. These statistics highlight the importance of the packaging industry in consuming new materials and emphasize the need for sustainable solutions to minimize the environmental impact of these materials. Sustainable packaging solutions can involve using recycled materials, creating biodegradable and compostable packaging, and applying circular economy principles. These principles focus on keeping packaging materials in use for as long as possible through methods like reuse, recycling, and remanufacturing [26].

The development of new packaging materials and designs is crucial for minimizing the environmental effects of packaging. For instance, progress in bioplastics, derived from renewable sources and capable of being composted, provides a promising substitute for traditional plastics. Additionally, lightweighting, which involves reducing the amount of material used in packaging, can greatly decrease resource consumption and waste generation [27]. Policy measures are essential for driving the shift to sustainable packaging. Regulations mandating recycled content, prohibiting certain types of single-use plastics, and implementing extended producer responsibility schemes can encourage companies to adopt more sustainable practices. For example, the European Union's Circular Economy Action Plan includes measures to reduce packaging waste and promote the use of recycled materials [28]. The high use of virgin materials and the growth of packaging waste present significant environmental challenges. The packaging industry's reliance on new materials drives resource extraction and pollution, while inadequate waste management systems contribute to rising waste levels and environmental degradation. Addressing these issues requires a multifaceted approach that includes technological innovation, regulatory measures, and changes in consumer behavior. By adopting sustainable packaging solutions and implementing effective waste management practices, the environmental impact of packaging can be significantly reduced [29].

2.2 Complexity in recycling multi-material packaging

Multi-material packaging, often chosen for its durability and protective properties, presents a significant challenge in the recycling process. This type of packaging combines various materials, such as plastics, aluminum, and paper, to make use of the unique strengths of each material. While these combinations can improve product protection, prolong shelf life, and decrease overpackaging, they make the recycling process more complex because of the challenge of separating the materials. Recycling multi-material packaging involves several steps: collection, sorting, processing, and finally, converting waste into new products. The complexity arises primarily during the sorting and processing stages. Traditional recycling systems are designed to handle single-material streams efficiently, such as pure plastics, paper, or metals. However, multi-material packaging requires advanced technologies to separate the different components effectively.

Sorting issues: Multi-material packaging often passes through automated sorting facilities that are not equipped to distinguish and separate complex materials. These facilities rely on visual and mechanical sorting techniques, which can struggle to identify and segregate multilayered or fused materials [30].

Processing complications: Once collected, multi-material packaging must be processed to separate its constituent materials. This process can be labor-intensive and costly, involving multiple steps such as shredding, washing, and chemical treatments. Each material must be isolated in a form pure enough to be reused, which is often not feasible on a large scale [31].

Despite the challenges in recycling, multi-material packaging offers significant benefits. One of the main advantages is that it optimizes space and reduces overpackaging. By combining materials, manufacturers can create packaging that is both lightweight and strong, thus reducing the need for additional packaging layers.

Space optimization: Multi-material packaging can be designed to be more compact and efficient, minimizing the volume of packaging used per product. This not only reduces the overall material usage but also lowers transportation costs and emissions due to more efficient packing and shipping [32].

Overpackaging reduction: The use of multiple materials can enhance the protective qualities of packaging, allowing for thinner and fewer layers without compromising product safety. This approach helps to address the issue of overpackaging, where excessive material is used to protect products during transit and storage [33].

To address the challenges of recycling, various innovations and policy measures are being explored. One solution being investigated is advanced recycling technologies like chemical recycling, which can break down materials at the molecular level, making it possible to recover high-quality raw materials from multi-material packaging. This method is more effective at handling the complex structures of multi-material packaging [34]. Manufacturers are being encouraged to design recyclable packaging, such as using mono-materials where possible or creating multi-material packaging that can be easily disassembled into recyclable components. Material science innovations are also leading to the development of new materials that mimic the properties of traditional multi-material packaging while being easier to recycle [31]. The regulations are implementing policies to promote sustainable packaging practices, including setting targets for recycled content, banning certain types of nonrecyclable packaging, and encouraging extended producer responsibility schemes where manufacturers are accountable for the end-of-life management of their products [35]. The recycling of multi-material packaging is complex and presents significant challenges, but it also inspires innovation toward more sustainable solutions. Although multi-material packaging can help optimize space and reduce overpackaging, the recycling process requires advancements in technology and supportive regulatory frameworks.

2.3 Harmonization of labeling

The harmonization of packaging labeling in the European Union (EU) is an important initiative aimed at standardizing labeling requirements across member states. This effort is driven by the need to reduce complexities and inconsistencies stemming from the various national labeling regulations in place. The European Commission has proposed a regulation on packaging and packaging waste (PPWR) to address these challenges by advocating for standardized labeling practices.

The current framework allows member states to create their own labeling systems, which results in a disorganized regulatory environment. This causes operational inefficiencies for companies that operate in multiple countries. For example, products have to be relabeled to meet each country's specific requirements, leading to logistical and financial challenges. This process not only requires additional labeling but also necessitates shipping products back to central logistics centers for compliance before they can be distributed to their final destinations. These practices pose difficulties for small and medium-sized enterprises that may not have the resources to consistently adapt to varying national regulations. The absence of standardized labeling also has environmental consequences. Packaging that is kept in stock may become noncompliant with new regulations and therefore needs to be disposed of, adding to waste. Additionally, varying national labeling requirements result in higher greenhouse gas emissions because of the extra transportation and handling needed for relabeling processes. The industry has expressed its support for a unified approach, stating that it would improve the functioning of the internal market and help achieve the EU's Green Deal goals. A collective industry statement underscores the importance of a standardized labeling system to minimize operational inefficiencies and environmental impacts. The statement emphasizes that current national labeling practices hinder the single market, raise costs for consumers, and result in unnecessary waste. The alignment of labeling regulations offers numerous advantages, such as increased market efficiency, lower operational costs, and better environmental sustainability. A standardized labeling system would eliminate the need for multiple relabeling processes, reducing logistical complexities and the corresponding greenhouse gas emissions. Additionally, it would enable manufacturers to plan and procure packaging materials more effectively, leading to less waste and greater sustainability. Furthermore, standardized labeling can enhance consumer understanding and compliance with recycling practices. Uniform labels can provide clear and consistent information about the disposal and recycling of packaging materials, promoting responsible consumer behavior and supporting circular economy initiatives.

2.4 Overpackaging and Internet

The use of e-commerce is rapidly increasing, and in 2019, it made up 14.1% of total retail sales worldwide. In 2022, more than 131 billion e-commerce packages were shipped within and across the world's 13 major markets [22]. This, combined with the expectation of affordable or even free deliveries and short delivery times, has put pressure on supply chains. These expectations often lead to inefficient use of logistics and transport resources in supply chains, particularly in e-commerce. Packaging is a significant concern for businesses involved in e-commerce in business-to-business markets. Overpackaging leads to excessive use of energy and materials, increasing the environmental impact of transportation and production processes [36]. Therefore, packaging is an important issue for actors in e-commerce. Overall, packaging has been a major concern for industrial marketers for a long time, as Easton has noted, "the cost of packaging is an integral part of the costs of getting the product to the customer in good condition" [37]. The rise of e-commerce has revolutionized the retail landscape, providing consumers with unprecedented convenience and accessibility.

However, this shift has also led to significant environmental challenges, particularly in the form of overpackaging. Overpackaging in e-commerce refers to the use of excessive or unnecessary packaging materials to protect products during shipping. This practice not only contributes to waste generation but also exacerbates environmental pollution.

To mitigate the environmental impact of overpackaging, Amazon has implemented several initiatives aimed at reducing packaging waste and promoting sustainability. These initiatives include Frustration-Free Packaging (FFP), Ship in Own Container (SIOC), and various packaging certifications.

- The Frustration-Free Packaging (FFP) program, which Amazon launched in 2008, aims to minimize excessive packaging by creating easy-to-open packages made from 100% recyclable materials. These packages are also free from unnecessary components such as plastic bindings and wire ties. This initiative not only reduces waste but also improves the customer experience by eliminating frustrating packaging elements.

- The Ship in Own Container (SIOC) program allows products to be shipped in their original packaging without additional Amazon-branded boxes. This approach significantly reduces the amount of packaging material used and minimizes the environmental impact of each shipment. SIOC is particularly effective for larger items that do not require extra protection beyond their original packaging.

Reducing overpackaging has several benefits, including lowering material costs, reducing waste, and minimizing environmental impact. However, achieving these benefits poses several challenges. One major challenge is finding a balance between the need for product protection and the goal of using fewer packaging materials. In addition, there is a need for standardized guidelines and regulations to ensure consistent packaging practices across different regions and industries.

The issue of overpackaging in e-commerce is complex and requires a comprehensive approach. Initiatives such as Amazon's packaging standards show that it is possible to reduce packaging waste while still ensuring product protection and improving customer satisfaction. Regulatory measures, such as the EU's PPWR, play a crucial role in driving industry-wide changes and promoting sustainable packaging practices. By addressing the challenges of overpackaging through innovation, regulation, and consumer awareness, the e-commerce industry can significantly reduce its environmental footprint.

3. Strategies for plastics and packaging

3.1 The plastic mix mechanical recycling and chemical

Plastic waste management is now a vital aspect of the global sustainability agenda, particularly under the European Strategy for Plastics in a Circular Economy. The strategy highlights the importance of changing the way plastic products are designed, manufactured, used, and recycled to minimize environmental impact and enhance resource efficiency [38]. In addressing the increasingly problematic issue of plastic waste, the packaging industry utilizes two key strategies: mechanical and chemical recycling. Each method has unique benefits and specific challenges, making them complementary in a comprehensive approach to managing plastic waste.

Mechanical recycling has several advantages. It generally requires less energy than chemical recycling, making it a more sustainable option for clean and sorted plastic waste [20]. The economic viability of mechanical recycling is well-established, especially for high-quality plastic waste streams [34]. Moreover, by reprocessing plastics without altering their chemical structure, mechanical recycling minimizes environmental emissions and resource use compared to virgin plastic production [24]. However, mechanical recycling also encounters significant challenges. Quality degradation is a major issue, as thermal and mechanical stresses during processing often lead to a deterioration in the quality of recycled plastics, limiting their recyclability [39]. Mixed and contaminated plastic waste streams pose additional challenges, requiring extensive sorting and cleaning, which increases processing costs and energy use [40]. Additionally, some plastics, such as thermoset plastics and multilayer composites, are not suitable for mechanical recycling due to their complex structures [38].

On the other hand, chemical recycling, also known as feedstock recycling, involves breaking down plastic waste into its fundamental chemical constituents, which can then be used to produce new plastics or other chemicals. This process includes methods, such as pyrolysis, gasification, hydrocracking, and depolymerization. Chemical recycling can handle mixed and contaminated plastic waste that mechanical recycling cannot effectively process [39]. By converting plastics back to their monomers or other chemical feedstocks, chemical recycling can produce high-quality materials comparable to virgin plastics [30]. Furthermore, chemical recycling is versatile and can be applied to a broader range of plastics, including those not suitable for mechanical recycling, such as thermoset plastics and complex composites [39]. Although chemical recycling offers several advantages, it also presents some challenges. One drawback is that chemical recycling processes usually require more energy compared to mechanical recycling, which can partially offset the environmental benefits [41]. Furthermore, the economic viability of chemical recycling is currently less favorable than that of mechanical recycling, largely due to high operational costs and the need for advanced technologies [42]. Additionally, many chemical recycling technologies are still in the developmental or pilot stages, requiring further research and scaling before they can become commercially viable [43].

Combining mechanical and chemical recycling can improve the overall efficiency and effectiveness of plastic waste management. Mechanical recycling is suitable for high-quality, sorted plastics, while chemical recycling can handle more challenging waste streams. This integrated approach not only maximizes resource recovery but also reduces the volume of plastics sent to landfills or incineration, aligning with circular economy principles.

3.2 The biodegradability strategy

Biodegradability is an effective way to manage plastic waste, particularly when recycling is not feasible. Biodegradable plastics are designed to break down more quickly and completely than traditional plastics in specific environmental conditions.

Biodegradable polymers are polymers that can be broken down by biological activity, resulting in a decrease in their molar mass [44]. These polymers can be of various types: (i) biomass polymers (e.g., starch or cellulose); (ii) polymers produced by microorganisms, such as polyhydroxyalkanoates (PHAs); (iii) polymers chemically synthesized from monomers derived from agricultural resources, like polylactic acid (PLA); and (iv) polymers obtained from fossil resources, such as polypropylene (PCL) [45]. The biodegradation process of biodegradable polymers generally consists of four steps [46]:

i. The breakdown of the polymer into smaller particles;

ii. The depolymerization of the polymer chain into oligomers, dimers, or monomers;

iii. Bio-assimilation, which involves the uptake of the small molecules into microbial cells and the subsequent production of primary and secondary metabolites;

iv. Mineralization of the metabolites, resulting in the formation of organic materials and gases such as carbon dioxide, methane, water, and nitrogen.

Despite their biodegradable properties, in 2023, biodegradable polymers only make up about 0.8% of the 367 million tons of plastic produced worldwide annually [47]. However, biodegradable polymers account for about 40% of the world's $460 billion packaging industry, as reported by NERAC. The EU holds 59% of the world market in polymers, while North America holds 22% and Asia holds 19%. The increasing number of patents filed globally for biodegradable packaging indicates a growing market and industry interest in sustainable packaging solutions. According to the NERAC report, there are about 8843 patent publications related to biodegradable polymers, and around 6972 patent publications are specific to biodegradable packaging. According to literature studies, 95% of patents filed are primarily for pharmaceuticals and medical use [44].

Biodegradable plastics offer several advantages. They can help reduce the environmental impact of plastic waste by breaking down more quickly than traditional plastics, which reduces litter and marine pollution. Some biodegradable plastics can be composted in industrial facilities, converting waste into valuable compost that enhances soil quality [48]. Additionally, bio-based biodegradable plastics utilize renewable resources, reducing reliance on fossil fuels and decreasing greenhouse gas emissions associated with plastic production [49].

However, biodegradable plastics also face challenges. Many require specific conditions to degrade effectively, such as high temperatures in industrial composting facilities, which may not be available everywhere [46]. Biodegradable plastics can be more costly to produce than traditional plastics and may have different performance characteristics, such as lower durability or higher permeability, which can limit their applications [50]. There is also a concern that mismanagement of biodegradable plastics could lead to incomplete degradation in natural environments, causing potential environmental harm similar to conventional plastics [51]. To effectively reduce plastic pollution, biodegradable plastics need to be integrated into existing waste management systems. This integration involves developing infrastructure, such as industrial composting facilities, to handle biodegradable plastics effectively [52]. Clear labeling and public education are essential to ensure proper disposal methods and prevent contamination of recycling streams [48]. Implementing policies and regulations supporting the development and use of biodegradable plastics, such as incentives for bio-based materials and standards for biodegradability, is also crucial [28]. Future advancements in biodegradable plastics rely on enhancing their performance, reducing costs, and ensuring they work well with existing waste management practices.

Innovations in material science, such as developing new biodegradable polymers with improved properties and using biotechnological methods to create more efficient degradation pathways, are crucial for progress in this field [46]. A systems approach that considers the entire lifecycle of biodegradable plastics, from production and use to disposal and degradation, is essential for maximizing their environmental benefits and minimizing unintended consequences [51].

The management of plastic waste in the packaging industry involves mechanical and chemical recycling, as well as the creation of biodegradable plastics. Each approach has its own advantages and drawbacks, and it is important to have a comprehensive strategy that includes all of these methods to effectively tackle the issue of plastic pollution. By improving recycling technologies, enhancing the quality of biodegradable plastics, and promoting collaboration throughout the supply chain, the packaging industry can make a significant contribution to a more sustainable and circular economy.

4. Conclusions

The packaging industry is currently at a critical juncture where the imperative of sustainability intersects with the complexities of modern logistics and consumer behavior. As the global economy continues to expand, the demand for packaging is also expected to grow by 4% annually from 2021 to 2026, driven by the increased use of materials such as cardboard. This growth presents significant environmental challenges, emphasizing the need to embrace more sustainable practices and innovations within the sector. Traditional packaging materials and processes have a substantial environmental footprint, contributing significantly to greenhouse gas emissions and resource depletion. The extraction and processing of virgin raw materials, including plastics, paper, metals, and glass, not only consumes significant amounts of energy but also results in pollution and habitat destruction. Moreover, packaging waste forms a substantial portion of municipal solid waste, thereby adding to the burden on landfills and contributing to ocean pollution. The shift to a circular economy in the packaging sector is crucial to addressing these impacts. This entails the design of packaging that can be reused, recycled, or composted, keeping the materials in use for as long as possible and minimizing waste. Technological advancements in materials science, such as the development of biodegradable plastics, offer promising steps toward achieving these objectives.

Regulation plays a crucial role in guiding the industry toward sustainability. For instance, the EU's Action Plan for the Circular Economy has set ambitious targets to reduce packaging waste and promote the use of recycled materials. These regulations aim to encourage companies to innovate and adopt more sustainable practices, such as using recycled content and reducing the amount of packaging used. Extended Producer Responsibility (EPR) schemes further compel manufacturers to take responsibility for the entire life cycle of their products, from design to disposal. The packaging industry has seen several innovative approaches to reduce its environmental footprint. For example, the Frustration-Free Packaging (FFP) and Ship in Own Container (SIOC) programs by Amazon aim to minimize packaging and reduce waste by using recyclable materials and shipping products in their original packaging. These initiatives not only reduce the environmental impact but also improve customer satisfaction by simplifying the process of opening packages. However, addressing overpackaging remains a significant challenge, especially in the e-commerce industry, where packaging must ensure product protection during shipping. Consumer behavior also plays a critical role; although there is a growing preference for sustainable packaging, actual purchasing decisions often depend on factors such as cost and convenience. Educational campaigns and clear labeling can help fill this gap by informing consumers about the environmental benefits of choosing sustainable packaging.

The packaging industry is making progress toward sustainability, but it still faces several challenges. The complexity of multi-packaging recycling, the high costs associated with developing and implementing new materials and technologies, and the need for consistent regulatory frameworks across regions are significant barriers. Moving forward, a multifaceted approach is required, including technological innovations, supportive regulatory measures, and changes in consumer behavior. Collaborative efforts between governments, businesses, and consumers are essential to push the industry toward more sustainable practices. By focusing on reducing the use of materials, improving recyclability, and adopting eco-friendly principles, the packaging industry can significantly reduce its environmental impact and contribute to a more sustainable future. In conclusion, the industry's journey toward sustainability is complex and challenging, but it also presents opportunities for innovation and improvement. Embracing a circular economy approach, adhering to strict regulatory frameworks, and meeting consumer preferences for sustainability can help the industry achieve significant environmental and economic benefits. Continued efforts to develop sustainable packaging solutions will be crucial in shaping a greener and more efficient global economy.

Author details

Enrico Maria Mosconi*, Andrea Colantoni and Mariarita Tarantino
University of Tuscia, Civitavecchia, Italy

*Address all correspondence to: enrico.mosconi@unitus.it

References

[1] Banaeian N, Mobli H, Nielsen IE, Omid M. Criteria definition and approaches in green supplier selection—A case study for raw material and packaging of food industry. Production & Manufacturing Research. 2015;**3**(1):149-168. DOI: 10.1080/21693277.2015.1016632

[2] Miao X, Magnier L, Mugge R. Switching to reuse? An exploration of consumers' perceptions and behaviour towards reusable packaging systems. Resources, Conservation and Recycling. 2023;**193**:106972. DOI: 10.1016/j.resconrec.2023.106972

[3] Gustavo JU, Pereira GM, Bond AJ, Viegas CV, Borchardt M. Drivers, opportunities and barriers for a retailer in the pursuit of more sustainable packaging redesign. Journal of Cleaner Production. 2018;**187**:18-28. DOI: 10.1016/J.JCLEPRO.2018.03.197

[4] Sastre RM, de Paula IC, Echeveste MES. A systematic literature review on packaging sustainability: Contents, opportunities, and guidelines. Sustainability (Switzerland). 2022;**14**(11):6727. DOI: 10.3390/SU14116727/S1

[5] Silva N, Pålsson H. Industrial packaging and its impact on sustainability and circular economy: A systematic literature review. Journal of Cleaner Production. 2022;**333**:130165. DOI: 10.1016/j.jclepro.2021.130165

[6] Packaging and Packaging Waste Regulation (PPWR). European Commission; 2024

[7] Rigamonti L, Ferreira S, Grosso M, Cunha Marques R. Economic-financial analysis of the Italian packaging waste management system from a local authority's perspective. Journal of Cleaner Production. 2014;87:533-541. DOI: 10.1016/j.jclepro.2014.10.069

[8] CONAI. Nel 2022 Riciclato il 71,5% dei Rifiuti di Imballaggio. Conai - Consorzio Nazionale Imballaggi. Roma, Italy. Available from: https://www.conai.org/notizie/nel-2022-riciclato-il-715-dei-rifiuti-di-imballaggio/ [Accessed: May 03, 2024]

[9] Economia Circolare—Dinamica e Gestione delle Organizzazioni—Commento alla Norma UNI/TS 11820:2022—Via Itri—la mia zavardella. Available from: https://viaitri.blog/2023/05/15/economia-circolare-dinamica-e-gestione-delle-organizzazioni-commento-alla-norma-uni-ts-118202022/ [Accessed: July 23, 2024]

[10] Reichert CL et al. Bio-based packaging: Materials, modifications, industrial applications and sustainability. Polymers (Basel). 2020;**12**(7). DOI: 10.3390/POLYM12071558

[11] Geueke B, Groh K, Muncke J. Food packaging in the circular economy: Overview of chemical safety aspects for commonly used materials. Journal of Cleaner Production. 2018;**193**:491-505. DOI: 10.1016/J.JCLEPRO.2018.05.005

[12] Lu Z, Hasselström L, Finnveden G, Johansson N. Cost-benefit analysis of two possible deposit-refund systems for reuse and recycling of plastic packaging in Sweden. Cleaner Waste Systems. 2022;**3**:100048. DOI: 10.1016/j.clwas.2022.100048

[13] Niero M. Implementation of the European Union's packaging and packaging waste regulation: A decision support framework combining quantitative environmental sustainability assessment methods and socio-technical approaches. Cleaner Waste Systems. 2023;**6**:100112. DOI: 10.1016/j.clwas.2023.100112

[14] Lifset R, Kalimo H, Jukka A, Kautto P, Miettinen M. Restoring the incentives for eco-design in extended producer responsibility: The challenges for eco-modulation. Waste Management. 2023;**168**:189-201. DOI: 10.1016/j.wasman.2023.05.033

[15] Georgakoudis ED, Tipi NS. An investigation into the issue of overpackaging-examining the case of paper packaging. International Journal of Sustainable Engineering. 2020;**14**(4)590-599. DOI: 10.1080/19397038.2020.1780337. Available from: https://www.tandfonline.com/loi/tsue20

[16] Sumter D, de Koning J, Bakker C, Balkenende R. Key Competencies for Design in a Circular Economy: Exploring Gaps in Design Knowledge and Skills for a Circular Economy. Sustainability. 2021;**13**:776. DOI: 10.3390/su13020776

[17] Potting J, Hekkert MP, Worrell E, Hanemaaijer A. Circular Economy: Measuring Innovation in the Product Chain Policy Report. Netherlands: Netherlands Environmental Assessment Agency; 2017

[18] The Circular Economy for Plastics – A European Analysis. Plastics Europe Report; 2024

[19] Confederation of European Paper Industries. Thomas Nowak - EHPA, Annita Westenbroek – Cepi; 2023

[20] Hopewell J, Dvorak R, Kosior E. Plastics recycling: Challenges and opportunities. Philosophical Transactions of the Royal Society, B: Biological Sciences. 2009;**364**(1526):2115. DOI: 10.1098/RSTB.2008.0311

[21] Tencati A, Pogutz S, Moda B, Brambilla M, Cacia C. Prevention policies addressing packaging and packaging waste: Some emerging trends. Waste Management. 2016;**56**:35-45. DOI: 10.1016/J.WASMAN.2016.06.025

[22] Statistics Explained. Available from: https://ec.europa.eu/eurostat/statistics-explained/index.php?title=Packaging_waste_statistics&action=statexp-seat&lang=it [Accessed: July 22, 2024]

[23] Packaging Waste Statistics. Statistics Explained. Available from: https://ec.europa.eu/eurostat/statistics-explained/index.php?title=Packaging_waste_statistics [Accessed: July 22, 2024]

[24] Geyer R, Jambeck JR, Law KL. Production, use, and fate of all plastics ever made. Science Advances. Polymers. 2020;**12**(7):1558. DOI: 10.3390/polym12071558

[25] Jambeck JR et al. Plastic waste inputs from land into the ocean. Science (1979). 2015;**347**(6223):768-771. DOI: 10.1126/SCIENCE.1260352/SUPPL_FILE/JAMBECK.SM.PDF

[26] MacArthur E. Circular Economy Towards the Opportunities for the Consumer Goods Sector. Vol. 2. Chicago: Founding Partners of the Ellen MacArthur Foundation; 2013. Pre print version

[27] Wojnowska-Baryła I, Kulikowska D, Bernat K. Effect of bio-based products on waste management. Sustainability.

2020;**12**(5):2088. DOI: 10.3390/SU12052088

[28] European Commission. Circular Economy Action Plan. Bruxelles: European Commission. Available from: https://environment.ec.europa.eu/strategy/circular-economy-action-plan_en [Accessed: July 22, 2024]

[29] Coelho PM, Corona B, ten Klooster R, Worrell E. Sustainability of reusable packaging—Current situation and trends. Resources, Conservation and Recycling: X. 2020;**6**:100037. DOI: 10.1016/J.RCRX.2020.100037

[30] de Mello Soares CT, Ek M, Östmark E, Gällstedt M, Karlsson S. Recycling of multi-material multilayer plastic packaging: Current trends and future scenarios. Resources, Conservation and Recycling. 2022;**176**:105905. DOI: 10.1016/J.RESCONREC.2021.105905

[31] Ajwani-Ramchandani R, Figueira S, Torres De Oliveira R, Jha S, Ramchandani A, Schuricht L. Towards a circular economy for packaging waste by using new technologies: The case of large multinationals in emerging economies. Journal of Cleaner Production. 2020;**281**(25):125-139. DOI: 10.1016/j.jclepro.2020.125139

[32] Monnot E, Reniou F, Parguel B, Elgaaied-Gambier L. 'Thinking outside the packaging box': Should brands consider store shelf context when eliminating overpackaging? Journal of Business Ethics. 2019;**154**(2):355-370. DOI: 10.1007/s10551-017-3439-0

[33] Elgaaïed-Gambier L. Who buys overpackaged grocery products and why? Understanding consumers' reactions to overpackaging in the food sector. Journal of Business Ethics. 2016;**135**(4):683-698. DOI: 10.1007/s10551-014-2491-2

[34] Ragaert K et al. Design from recycling: A complex mixed plastic waste case study. Resources, Conservation and Recycling. 2019;**155**:104646. DOI: 10.1016/j.resconrec.2019.104646

[35] Joltreau E. Extended producer responsibility, packaging waste reduction and eco-design. Environmental and Resource Economics (Dordrecht). 2022;**83**(3):527-578. DOI: 10.1007/S10640-022-00696-9

[36] Ibrahim ID et al. Need for sustainable packaging: An overview. Polymers. 2022;**14**(20):4430. DOI: 10.3390/polym14204430

[37] Easton A. Purposes and analysis of industrial package design. Industrial Marketing Management. 1978;7(5):316-323. DOI: 10.1016/0019-8501(78)90073-1

[38] Davidson MG, Furlong RA, McManus MC. Developments in the life cycle assessment of chemical recycling of plastic waste—A review. Journal of Cleaner Production. 2021;**293**:126163. DOI: 10.1016/j.jclepro.2021.126163

[39] Ragaert K, Delva L, Van Geem K. Mechanical and chemical recycling of solid plastic waste. Waste Management. 2017;**69**:24-58. DOI: 10.1016/J.WASMAN.2017.07.044

[40] Hahladakis JN, Velis CA, Weber R, Iacovidou E, Purnell P. An overview of chemical additives present in plastics: Migration, release, fate and environmental impact during their use, disposal and recycling. Journal of Hazardous Materials. 2018;**344**:179-199. DOI: 10.1016/J.JHAZMAT.2017.10.014

[41] Akah A, Hernandez-Martinez J, Rallan C, Garforth A. Enhanced feedstock recycling of post-consumer

plastic waste. Chemical Engineering Transactions. 2015

[42] Maga D, Hiebel M, Thonemann N. Life cycle assessment of recycling options for polylactic acid. Resources, Conservation and Recycling. 2019;**149**:86-96. DOI: 10.1016/J.RESCONREC.2019.05.018

[43] Munir D, Irfan MF, Usman MR. Hydrocracking of virgin and waste plastics: A detailed review. Renewable and Sustainable Energy Reviews. 2018;**90**:490-515. DOI: 10.1016/J.RSER.2018.03.034

[44] Lee CJ, Ma HR, Kim YT. Patent status of biodegradable polymers and identification of new application areas by IPC network analysis. World Patent Information. 2024;**78**:102298. DOI: 10.1016/j.wpi.2024.102298

[45] Avérous L, Pollet E. Biodegradable polymers. Environmental Silicate Nano-Biocomposites, Chapter 2, nergy and Technology. 2012;**50**:13-39. DOI: 10.1007/978-1-4471-4108-2_2

[46] Haider TP, Völker C, Kramm J, Landfester K, Wurm FR. Plastics of the future? The impact of biodegradable polymers on the environment and on society. Angewandte Chemie, International Edition. 2019;**58**(1):50-62. DOI: 10.1002/anie.201805766

[47] Global Production Capacities of Bioplastics 2021-2026. Global Production Capacities of Bioplastics 2021 (By Material Type). Global Production Capacities of Bioplastics 2026 (By Material Type). Available from: http://www.european-bioplastics.org/news/publications/ [Accessed: August 28, 2024]

[48] Dilkes-Hoffman L, Ashworth P, Laycock B, Pratt S, Lant P. Public attitudes towards bioplastics—Knowledge, perception and end-of-life management. Resources, Conservation and Recycling. 2019;**151**:104479. DOI: 10.1016/J.RESCONREC.2019.104479

[49] Engineering S et al. Greenhouse gas mitigation for U.S. plastics production: Energy first, feedstocks later. Environmental Research Letters. 2017;**12**(3):034024. DOI: 10.1088/1748-9326/AA60A7

[50] Rujnić-Sokele M, Pilipović A. Challenges and opportunities of biodegradable plastics: A mini review. Waste Management & Research. 2017;**35**(2):132-140. DOI: 10.1177/0734242X16683272

[51] Narancic T et al. Biodegradable plastic blends create new possibilities for end-of-life management of plastics but they are not a panacea for plastic pollution. Environmental Science & Technology. 2018;**52**(18):10441-10452. DOI: 10.1021/ACS.EST.8B02963

[52] Global Production Capacities of Bioplastics 2018-2024. Global Production Capacities of Bioplastics 2019 (By Material Type). Global Production Capacities of Bioplastics 2024 (By Material Type). Available from: http://www.european-bioplastics.org/news/publications/

Chapter 11

Practice of Expired and Unused Drug Disposal in South Asian and Southeast Asian Countries

Mohammad Borhan Uddin, Ratin Al Samir and G.M. Golam Mehebub Pantho

Abstract

The rising incidence of diseases and subsequent medication use have led to an increase in unused, expired, and unwanted medications. Improper disposal practices, prevalent in South Asian countries due to the lack of official guidelines, pose significant threats to public health and the environment. Multiple studies were conducted during the last decade across South and Southeast Asian regions, involving thousands participants. The predominant disposal methods include discarding medications in household trash or flushing them down sinks/toilets, with minimal returns to pharmacies for proper disposal. Lack of awareness, home storage of medications, and inadequate disposal methods were common trends identified. Dosage forms, particularly topical preparations, significantly influenced disposal practices. The consequences of improper disposal on human health encompass environmental contamination, antibiotic resistance, and potential physiological disruptions, emphasizing the need for education, intervention, and regulatory measures. Environmental ramifications extend to water and soil contamination, necessitating eco-friendly disposal strategies and robust medication take-back programs. Preventive measures might include improved drug management systems, community "take-back" programs, and exploring donation options for unused medications. Collaborative efforts among stakeholders, emphasizing the role of pharmacists, are crucial in establishing safe disposal guidelines and addressing pharmaceutical waste management challenges.

Keywords: improper drug disposal, unused drugs, expired drugs, pharmacists, pharmaceutical waste management

1. Introduction

Since disease incidence and prevalence have been rising over the past few decades, especially from COVID-19, healthcare practitioners must prescribe and dispense numerous medications. Most of the time these prescribed medicines are left unused due to adverse effects, dosage adjustment, recovery from disease, expiration date, physician prescribing practice, dispenser practice, and so forth [1]. Unused, unwanted, and expired drugs kept at home endanger both health systems and the environment [2].

Bangladesh, like many South and South Asian developing nations, lacks clear government regulations for disposing of expired and unwanted pharmaceuticals [3]. For this reason, many hazardous methods such as throwing medications in the sink, trash, or toilet are common. This improper disposal negatively affects environmental health and safety [4]. For instance, expired tetracyclines can cause renal tubular damage [5]. The decline in vulture populations is linked to diclofenac exposure, and improper disposal also risks accidental poisoning and abuse, potentially contributing to antimicrobial resistance [5].

Drug disposal practices vary widely globally due to many reasons; most of them are due to government regulations [6–9], individual cultural attitudes [10–15], infrastructure of the healthcare system [16–20], and last but not the least public awareness [8, 21–25]. These factors can affect drug disposal methods in different countries including South and South Asian regions. Understanding these factors helps to appreciate why drug disposal practices are so varied around the world and highlights the importance of context when addressing pharmaceutical waste management. These key confounding factors have been discussed in short here:

Regulatory frameworks and economic factors: Based on the socioeconomic perspectives, globally, different countries have different rules and regulations governing the disposal of pharmaceuticals and personal care products. In developed countries, there are stringent guidelines and programs for proper disposal. Advances in technology and science in these developed countries can influence disposal methods. Some rich countries may use advanced techniques to neutralize or treat pharmaceutical waste. The strength of environmental regulations and policies in these countries can affect drug disposal practices. Countries with strict environmental laws may have more robust systems for handling pharmaceutical waste. On the other hand, less developed countries may rely on less effective methods [8]. In these regions, regulations may be less developed or poorly enforced. There may be illegal or improper disposal practices, such as flushing drugs down the toilet or throwing them in the trash, which can contribute to environmental and public health issues.

Infrastructure of healthcare system: The structure and efficiency of the healthcare system can influence greatly drug disposal. Countries with well-organized healthcare systems can practice better disposal programs and facilities [20]. On the other hand, countries with less developed systems might struggle with effective disposal. In some countries, there are designated take-back programs, drop-off locations, and community collection events, while others may lack such infrastructure [26]. The practices of pharmaceutical companies regarding take-back programs or disposal guidance can impact how drugs are disposed of [27].

Public awareness and education: Awareness about the environmental and health risks associated with improper drug disposal varies. In First World countries, public education campaigns are in place, while in others, there might be limited information available to the public [25].

Cultural attitudes: Cultural perceptions of drug use and disposal can influence practices. In some cultures, there might be a strong emphasis on environmental protection, leading to more conscientious disposal practices [15], whereas in others, there might be less awareness or concern.

Swift and immediate attention is imperative to tackle this serious issue, particularly in the South Asian region. Taking the necessary measures to address this issue would not only improve the population's standard of life but also contribute to the

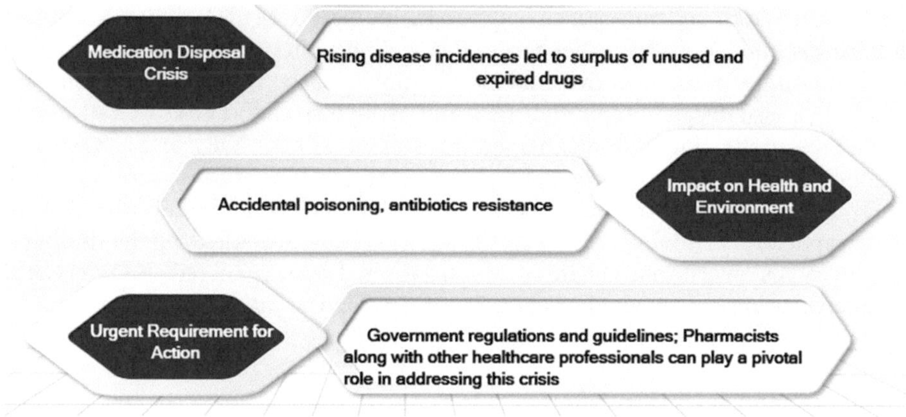

Figure 1.
Impact of improper drug disposal practice [28, 32].

worldwide battle to combat medication desensitization, such as antibiotic resistance. Pharmacists can significantly influence this matter by playing a crucial role. Proper disposal of collected medicines by community pharmacists can effectively mitigate the harmful consequences associated with medication disposal [28].

Numerous studies in recent years from different countries have evaluated the existing knowledge and practices regarding unused and expired medicines among healthcare workers. Researchers explored the disposal practices among university students [4, 29], healthcare professionals [30], and households [31]. To the best of our knowledge, very few comprehensive studies were conducted to assess the knowledge, attitude, and practice of unused and expired medications among the people in South and Southeast Asian Countries. The aim and objective (**Figure 1**) of this narrative review are to assess the current drug disposal practices mainly among students, patients, doctors, nurses, and pharmacist professionals in South and South Asian Countries, and to identify the most prevalent methods of drug disposal used by these healthcare practitioners. Subsequently, based on the findings, the study aims to propose a concrete and comprehensive guideline for safe and environmentally responsible drug disposal to promote adherence to proper disposal practices and reduce the potential adverse impacts on public health and the environment.

Literature shows a lack of availability of public data about the safe and effective disposal knowledge and practices toward expired medication. That is why this review has been planned to report the current knowledge, practices, and attitudes in these selected Asian countries and to focus on more comprehensive researches that are needed to be conducted more frequently to get more data about improper medicines disposal practices.

2. Method

2.1 Searches and strategies

A non-systematic narrative review was conducted to examine the knowledge on the concern of drug disposal procedures to provide better public awareness of the problem of incorrect disposal practices in South and Southeast Asian countries. Data were gathered from multiple search engines, including PubMed, Elsevier, Nature, and Springer Link, as well as other acceptable and relevant papers and guidelines, to satisfy our studies' aim, objective, and justification. The keywords for searching were used like "unused drugs," "expired drugs," and "drug disposal". The search strategy included the application of Boolean operators such as "AND," "OR," and "NOT," which can either limit or widen the scope of the search records. The studies were reviewed by analyzing research instruments like structured interviews and questionnaires from research articles, narrowing articles by fixing the focus and evaluating the findings based on exclusion criteria. The information gathered was organized and referenced to offer an overview to the readers about the drug disposal practices in South and Southeast Asian countries.

2.2 Inclusion and exclusion criteria

All of the research articles from which the information obtained had been published since 2010. Most of the studies involve questionnaire-based surveys related to drug disposal methods and procedures adopted by students, doctors, nurses, pharmacists and patients in South and South East Asian countries. These studies were focused on the practice of disposing of pharmaceuticals and public awareness of the problem of incorrect disposing methods. Articles were excluded if they did not focus specifically on drug disposal practices and conducted in other Asian countries (**Figure 2**).

The resultant 17 text records were screened for full-text eligibility; these records were found to be published from 2010 to 2023. These 17 records included in this analysis showed participants from diverse backgrounds (**Figure 3**).

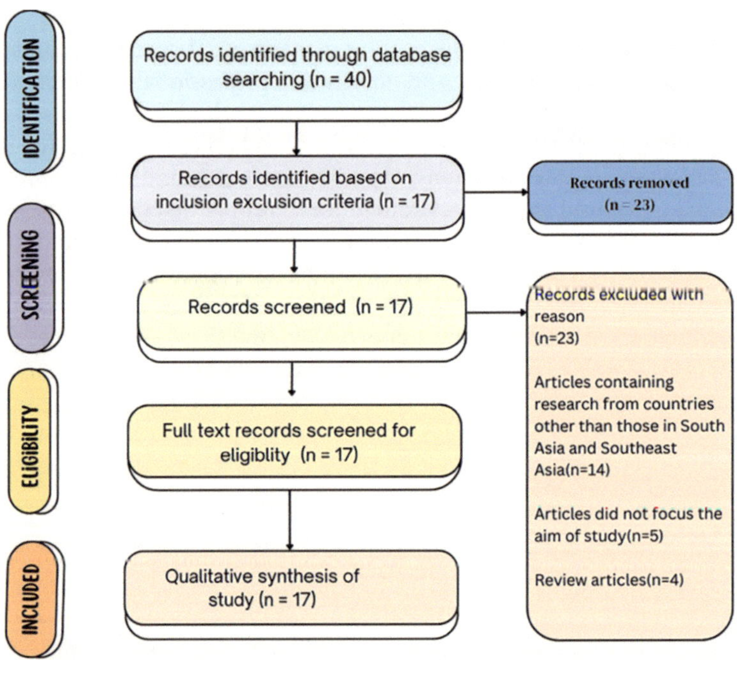

Figure 2.
Diagram of inclusion and exclusion criteria.

Figure 3.
Background of the studies' participants.

3. Result

It is critical to safely dispose of unused and expired pharmaceutical products to reduce environmental contamination, public health dangers, and potential drug misuse. This analysis (**Table 1**) will take a peek at how unneeded and expired pharmaceutical products are disposed of in South and Southeast Asia.

3.1 Conventional practices for drug disposal in different nationals

According to a review of multiple research studies, the most typical manner of disposing of drugs in South Asian nations is through the trash or dustbin. Secondly, the

Region	Study design/sample size	Observation	References
India	Questionnaires n = 118 (medical students)	The majority (63.5%) of medications that were stored were in topical preparations. About 73% of students said they would dispose of unused or expired prescriptions in dustbins at home.	[29]
India	Cross-sectional questionnaire-based study n = 400 (doctors, medical students, staff nurses, and pharmacy students)	About 17% of participants stated that crushing the drugs before discarding them in the trash is the primary method of drug disposal. Throwing away drugs (17%), flushing them into the toilet or sink (6%), and only 5% returning them to the pharmacy.	[24]
India	Questionnaire-based cross-sectional study n = 120 (medical and para medical students)	The two most common drug disposal methods consisted of dumping drugs in the trash and flushing them down the toilet. About 66% of respondents thought returning to the pharmacist was the most acceptable method of drug disposal.	[33]
India	Questionnaire-based cross-sectional study n = 956 (general patients)	The majority of consumers (87%) kept their medications at home. Almost all (92.6%) of the consumers threw away the expired medicines after storing them for a few days. Consumers discarded the expired medicines mainly in household trash (73%).	[34]

Region	Study design/sample size	Observation	References
South India	Questionnaires; face-to-face semi-structured interview n = 480 (nurses)	About 83.3% of the participants were ignorant of the negative environmental effects of drug disposal; most of them (45.97%) discarded unused medicines in the trash.	[30]
North India	Questionnaires (both open and close-ended questions) n = 236 (undergraduate dental students)	Most of the participants accredited their unawareness of the drug reactions to the environment where the main drug disposal system is trashing, which is 94% of the population.	[14]
Bangladesh	Structured questionnaires n = 250 (pharmacy and general students)	Undergraduate students in Bangladesh have alarmingly low knowledge and awareness about remaining household drug disposal. Participants have been identified to keep analgesics, antibiotics, vitamins, antihistamines, gastric medicines, and antipyretic drugs, which can create a threat to the public's health.	[4]
Bangladesh	Structured questionnaires n = 400 (general patients)	Without further processing, 47% of respondents threw expired medications in the trash, 19% dropped them out the window, 4% used the medication again, 2% burned them in a fire, 21% gave them to the pharmacy for a refund, and 4% flushed them down the toilet.	[35]
Malaysia	Interview n = 28 (patients)	The most common disposal method (57.1%) was throwing unused medication into the trash; 55% of participants were unaware of the adverse consequences of unused medication.	[36]
Malaysia	Structured questionnaires n = 885 (undergraduate students)	Although 87% of the population was aware of drug waste, only 2% used the drug take-back scheme. Additionally, 83% of participants threw away unneeded drugs in landfills.	[37]
Malaysia (Rawang)	Questionnaire (close-ended question) n = 384 (parents of school-aged children)	The majority of the participants were clueless about medicine waste, which is around 69.5%; most of them (74.3%) threw medicine waste into a waste bin.	[38]
Malaysia	Cross-sectional study n = 426 (general patients)	More than 80% of the participants in the study claimed they were aware of the growing issue of medication waste in Malaysia, its effects on patients and the economy, and its detrimental effects on human health and the environment.	[31]
Nepal	A web-based cross-sectional descriptive study n = 294 (healthcare professionals)	Most disposal methods implemented by healthcare professionals were not recommended. Healthcare professionals had an elevated likelihood of keeping medicinal products at home.	[2]

Region	Study design/sample size	Observation	References
Afghanistan	Face-to-face interviews using a pre-validated structure n = 301 (general patients)	Expired medicines were dumped in the trash by 77.7% of interviewees. A large percentage of the survey participants held the government accountable for raising awareness about proper medicine disposal. Almost all of those interviewed (98%) agreed that careless disposal of unwanted and expired medications can harm the ecosystem and human health.	[5]
Indonesia	Cross-sectional study n = 497 (general patients)	Throwing away unwanted medicines in household garbage was the most typical way of disposal which is around 82%.	[39]
Pakistan	Close-end question n = 1022(General patients)	Around 80% of the population stored unused medication in their home, while 20% completely consumed purchased medication	[40]
Thailand (Khonkaen City)	Structured interview n = 331 (general patients)	A majority of the participants (73%) had supported that drug take-back program could be used to manage the leftover drugs issues. However, most of them (89%) mentioned that they were not being taught about safe drug disposal.	[41]

Table 1.
Overview of drug disposal practices in different South and South Asian countries.

Method of disposal					
Region	**Dustin or trash**	**Sink or toilet**	**Other(s)**	**Returning to pharmacy**	**References**
Nepal	72%	21%	—	2%	[2]
Malaysia	84%	12%	—	5%	[31]
India	73%	20%	—	—	[34]
Pakistan	83%	12%	—	2%	[40]
Indonesia	82%	5%	4% burned the medicine	0.2%	[39]
Bangladesh	47%	—	2% burned the medicine	21%	[35]

Table 2.
Conventional methods of handling leftover/expired medication in different countries.

practice of flushing medications down sinks or toilets is also common in countries like India, Pakistan, Indonesia, Malaysia, and Bangladesh. Most critically, the tendency to return unused and expired medicine to pharmacies for appropriate disposal is rarely seen (**Table 2**).

4. Discussion

From **Table 1**, it can be said that in India, it was found that 63.5% of stored medications were topical preparations, with 73% of respondents preferring disposal in household dustbins. About 17% of healthcare personnel crushed medicines before throwing them away, which is terrifying, while just 5% returned them to the pharmacy. Additionally, a significant number of healthcare professionals kept medicines at home, employing non-recommended disposal methods.

Low knowledge and awareness of proper drug disposal were observed among pharmacy students and individuals in Bangladesh. Moreover, many respondents stored various drugs at home, posing potential health hazards. In Malaysia, over 80% of the population was aware of the growing issue of medication waste and its effects on patients and the circumstances. Disappointingly, common improper disposal techniques in India included throwing drugs in the trash or flushing them down the toilet. On a positive note, 66% preferred returning drugs to the pharmacist for proper disposal. Another study reported that 87% of consumers kept medications at home, with about 92.6% discarding expired medicines in household trash in India and 77.7% in Afghanistan. Interestingly, nearly the entire sample acknowledged the potential environmental and health impacts of improper disposal. Meanwhile, in Indonesia, 82.1% of students admitted to throwing away unwanted medicines in households, and approximately 80% of the population in Pakistan stored unused medication at home. The findings from these studies underscore some common trends and challenges in medication disposal practices across the studied countries:

Lack of awareness: A significant portion of participants exhibited insufficient knowledge and awareness of proper medication disposal methods, thereby posing environmental and health risks.

Home storage: A considerable number of respondents, including healthcare professionals and students, stored medications at home, resulting in possible medication wastage and unauthorized access.

Inadequate disposal methods: Many individuals have the attitude and practice of disposing of the drugs in household trash, flushing them down the toilet, or even burning them, which has negative environmental repercussions.

Need for education and intervention: There is an urgent need for educational campaigns and interventions to enhance awareness regarding proper medication disposal among healthcare professionals, students, and the general population.

The research data from various South Asian and Southeast Asian countries accentuate the importance of addressing medication disposal practices to safeguard public health and the environment. Encouraging the adoption of proper disposal methods and raising awareness among healthcare professionals, students, and the general population are critical steps to mitigate the risks associated with improper medication disposal. Collaborative efforts among government agencies, healthcare institutions, and educational establishments are essential in devising and implementing effective strategies for safe and eco-friendly medication disposal.

4.1 Impact of dosage forms on drug disposal practices

The influence of dosage forms on medicine disposal habits is evident from a comprehensive analysis of recent studies. Topical formulations emerged as the most stored drug group (63.5%), followed by antipyretics (16.9%), antibiotics (13.5%), antacids (2.5%), analgesics (2.5%), and antihistamines (1.6%) [2]. This contrasts with a different study where analgesics (34%) were the predominant stored drugs, followed by antibiotics (29%) and antipyretics (11%). Regional prescription trends and disease prevalence may account for these variations. The study involved 120 participants, including paramedical (68) and medical (52) students. A majority

(53%) reported having 1–5 unused medicines, 29% had 6–10, and 14% had 11–15. Commonly stocked medicines included topical drugs (40%), analgesics and antipyretics (25%), antibiotics (24%), and vitamins and syrups (8%) [33]. Rational medicine purchase practices were reflected, with most respondents obtaining medicines on prescription. Interestingly, education played a vital role, as university graduates in other Asian countries primarily purchased prescription medicines (142 students), while a few (19 students) obtained over-the-counter drugs. Awareness of expiration dates was high (95.5%) among respondents, contributing to prudent purchase decisions [5]. Unused medicines were attributed to improved medical conditions (82.7%), expiration (6.2%), prescription alterations (5%), nonadherence (1.8%), adverse events (1.6%), and uncertainty about medication purpose (1.0%). NSAIDs accounted for the most underutilized medications (372) after vitamins and minerals then nutritional supplements (215), and antimicrobial agents (171). This underscores the importance of considering dosage forms in understanding and promoting responsible drug disposal practices [39].

4.2 Consequences of improper drug disposal on human health

The consequences of inadequate medication disposal pose a grave threat to human health, and recent research has shed light on the intricate pathophysiological mechanisms underlying this issue. Inappropriate disposal practices, including flushing medications down the toilet or discarding them in household waste, contribute to the contamination of water sources with pharmaceutical residues [42]. The presence of these active pharmaceutical ingredients in water systems has been associated with a myriad of adverse health effects. Notably, the interference with endocrine systems due to chronic exposure to low concentrations of pharmaceuticals has been implicated in disruptions to hormonal regulation and potential developmental abnormalities [42]. Furthermore, the rising incidence of antibiotic resistance, caused by pharmaceutical products released into the environment, represents a serious risk to the safety of humans by endangering the efficacy of necessary medicinal treatment options [32]. The intricate interplay between pharmaceutical residues and human physiology, as elucidated in recent research, underscores the critical importance of adopting proper medication disposal practices to mitigate the pathophysiological consequences and safeguard the well-being of individuals.

Moreover, the impact of improper medication disposal extends beyond water contamination, affecting systemic health through complex biological pathways. Highlighting the potential for bioaccumulation of pharmaceutical residues in the food chain, recent studies have drawn attention to this environmental concern. This introduces an additional layer of concern, as humans may be exposed to these contaminants by consuming contaminated food products [42]. The bioaccumulation of pharmaceuticals may result in cumulative exposure, amplifying the potential for adverse health effects [43]. The intricate interplay between medication residues and human physiology underscores the need for a comprehensive understanding of the pathophysiological consequences of inadequate disposal. Additionally, the persistence of certain pharmaceutical compounds in the environment raises concerns about chronic exposure and the long-term impact on human health. These findings collectively emphasize the urgent necessity for targeted interventions, public awareness campaigns, and regulatory measures to address the pathophysiological implications of inadequate medication disposal and mitigate the associated risks to human health. Furthermore, the consequences of inadequate medication disposal extend beyond the direct physiological impacts and delve into the societal and public health dimensions. Improper drug disposal practices play a big role in contributing to the broader issue of substance misuse and abuse of over the counter and prescription drugs. The presence of unwanted or expired prescriptions in households, as a result of improper disposal approaches, may contribute to non-prescribed drug use, particularly opioid and other controlled drug abuse [44]. This societal aspect of inadequate disposal adds a layer of complexity to the pathophysiological consequences, intertwining individual health risks with the broader public health crisis of substance abuse. Inappropriate disposal methods contribute to the contamination of water sources with active pharmaceutical ingredients (APIs), setting the stage for potential disruptions in endocrine function and the development of antibiotic-resistant bacteria [45]. These phenomena have direct implications for the pathophysiology of various human diseases, with disruptions in hormonal regulation linked to conditions such as reproductive disorders and developmental abnormalities [46]. Furthermore, the surge in antibiotic resistance, fueled by pharmaceutical residues in the environment, poses a significant threat to the efficacy of medical treatments for infectious diseases, exacerbating the global health challenge of antimicrobial resistance [32]. The intricate interplay between improper medication disposal and the pathophysiology of human diseases accentuates the critical need for targeted interventions, public awareness campaigns, and regulatory measures to mitigate these health risks.

In **Figure 4**, we can see the effects of improper disposal of antibiotics in Bangladesh. It emphasizes that improper disposal significantly contributes to antibiotic resistance.

4.3 Environmental threats of improper drug disposal practices

Improper disposal of medications poses significant environmental threats, as API (Active Pharmaceutical Ingredients) & excipients can find their way into water sources, soil, and ecosystems, leading to detrimental consequences for both human and environmental health. The alarming persistence of pharmaceutical residues in

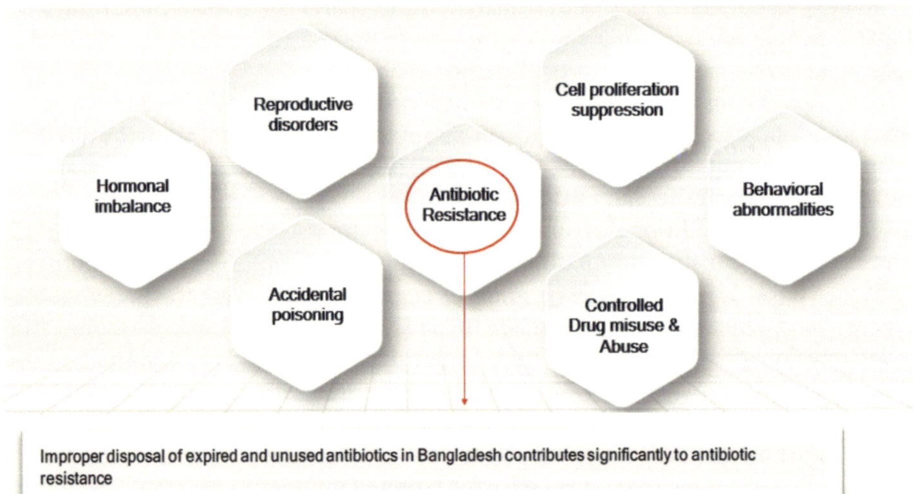

Figure 4.
Consequences of improper drug disposal on human health in Bangladesh [8, 32, 47].

the environment emphasizes the need for responsible medication disposal practices. Pharmaceutical compounds, even at low concentrations, have been detected in various environmental matrices, including surface water and sediments [32]. Such contamination can result in adverse effects on aquatic organisms, potentially disrupting ecosystems and leading to the development of antibiotic-resistant strains [48]. The improper disposal of unused or expired medications, whether through flushing them down the toilet or discarding them in household trash, has been identified as a source of pharmaceutical pollution, further accentuating the urgency of implementing effective medication take-back programs and raising public awareness. The environmental impact of pharmaceutical pollution extends beyond immediate aquatic ecosystems, with studies indicating the potential for bioaccumulation in higher trophic levels, including humans. This emphasizes the interconnectedness of environmental and human health and underscores the importance of adopting eco-friendly medication disposal practices. Underscoring the critical need for proactive measures to address the environmental ramifications of improper medication disposal, this emphasizes the importance of informed public engagement and the establishment of robust medication take-back systems to mitigate the risks posed by pharmaceutical pollution [32, 49]. Additionally, the issue of pharmaceutical contamination extends to agricultural systems [42]. The improper disposal of medications can lead to the introduction of pharmaceutical residues into soils through the application of contaminated water or sewage sludge as fertilizers. This introduces a new dimension to the environmental impact of improper medication disposal, as these residues can potentially affect crop growth and quality. Moreover, the presence of pharmaceutical compounds in soil raises concerns about their potential uptake by plants, which could result in the entry of these substances into the food chain, ultimately reaching humans.

4.4 Possible preventive measures of improper drug disposal

The issue of drugs reaching expiry dates before utilization in drug supply management is well known. Expired drug stocks pose significant problems, leading to financial losses and requiring safe disposal to prevent harm to people and the environment. Expired drugs necessitate safe disposal due to financial losses incurred when they can no longer be distributed. Additionally, pharmaceutical products may lose potency during the degradation process, resulting in reduced effectiveness. Manufacturers are responsible for any harmful effects within a product's shelf life, provided proper transportation and storage conditions are maintained. Establishing a robust drug management system is crucial, including drug estimation based on health service data, effective stock inventory control, adherence to "First Expiry First Out" (FEFO) and "First In First Out" (FIFO) principles, coordination with health institutions, and negotiation with suppliers for potential returns of soon-to-expire drugs [50]. Moreover, some other possible preventive measures can be taken (**Figure 5**). Implementing these steps can reduce future environment ramifications.

4.4.1 Drug take-back program

The improper disposal of pharmaceuticals, by either flushing them down the toilet or throwing them in the trash, has led to a significant increase in pharmaceuticals in water sources. Returning expired drugs through community drug "take-back" programs would be an ideal solution, but in many Southeast Asian nations, these are not generally available, restricting people's disposal alternatives. A study involving

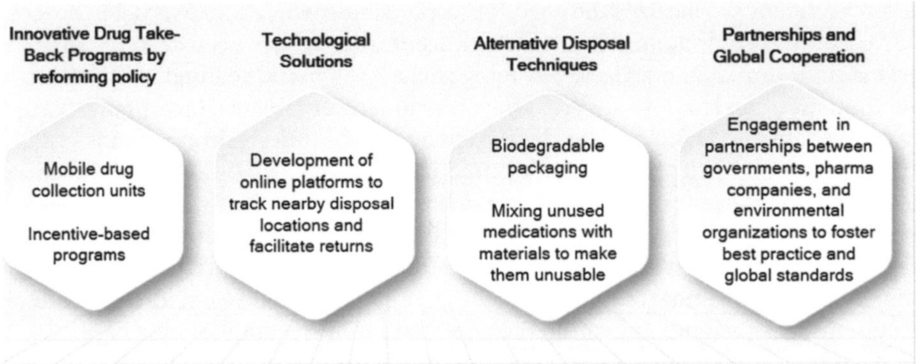

Figure 5.
Possible preventive measures of improper drug disposal [4, 28].

community pharmacists revealed that a majority of respondents recognized the potential of "drug take-back programs" to address improper medical disposal. The lack of standards on proper disposal practices among pharmacists, on the other hand, contributed to a problem, with roughly half of them rejecting to use pharmacies as medication take-back locations.

According to a survey held in India, 72% of pharmacists believe that returning unnecessary pharmaceutical products to government hospitals and institutions is the best alternative. The study also highlighted the possibility of donating unused medicines to those who cannot afford them as an alternative to throwing away expensive medications or letting them expire. Drug "take-back" systems are well-established in prosperous nations like the United States, and they make a substantial contribution to appropriate drug dispensing and disposal [51]. To address the issue, environmental practices such as incineration and disposal of pharmaceuticals in impermeable containers with refuse solids, like coffee grounds, should be implemented. This is especially important for unused and discarded drugs that involve analgesics, antibiotics, and medicines affecting the brain and spinal cord, including controlled medicines with dependent risks including opioids.

4.4.2 Technological solutions

Developing online platforms to track disposal locations and facilitate returns is a crucial technological solution for managing medication disposal. These platforms can provide users with real-time information on the nearest drug take-back locations, ensuring that expired or unused medications are disposed of properly. Additionally, they can offer guidance on how to return medications safely and efficiently. Such platforms can greatly reduce the hazards to health and the environment connected with incorrect pharmaceutical disposal by providing easy access to this information. Additionally, they can support the monitoring and analysis of disposal trends, which can help to enhance waste management techniques.

4.4.3 Alternative disposal techniques

Sustainable solutions for medicine waste include the use of biodegradable packaging and the combining of unwanted medications with ingredients to render them unusable. By preventing the packaging from adding to pollution, biodegradable packaging helps to lessen its negative effects on the environment. In the meanwhile, drug abuse and unintentional consumption are avoided by combining unwanted pharmaceuticals with substances that render them useless. By ensuring that the pharmaceuticals cannot be recovered and used after being disposed of, this procedure safeguards the public's health and safety. These methods reduce the possibility of harm from discarded drugs and encourage environmentally sustainable habits.

4.4.4 Partnerships and global cooperation

Encouraging partnerships between governments, pharmaceutical companies, and environmental organizations is essential for fostering best practices and global standards in medication disposal [52]. Such collaborations can lead to the development and implementation of comprehensive policies and programs that address the complexities of pharmaceutical waste management. By working together, these entities can share knowledge, resources, and technologies to create more effective disposal systems. Global cooperation ensures that best practices are adopted universally, helping to mitigate the environmental and health impacts of improper medication disposal on a larger scale. These partnerships also facilitate the alignment of regulations and standards, promoting a coordinated approach to tackling this global issue.

4.4.5 Limitations

Our study has some limitations. This chapter has focused on qualitative summaries rather than quantitative synthesis and also did not involve the statistical synthesis of the data. So, researchers and readers should be aware of these limitations when interpreting our findings.

5. Conclusion

In conclusion, the improper disposal of pharmaceuticals presents a significant challenge in South Asian and Southeast Asian countries. The absence of official guidelines for drug disposal has led to unsafe practices, raising considerable environmental and public health concerns. Urgent attention is needed to address these issues, emphasizing the crucial role of pharmacists and advocating for the establishment of community drug "take-back" programs. The proposed guideline for safe drug disposal, coupled with preventive measures, underscores the necessity for collaborative efforts among stakeholders. By highlighting the environmental and health risks associated with inadequate medication disposal, this study provides a basis for informed decision-making and proactive interventions. The hope is that these insights will drive positive change, promoting responsible pharmaceutical waste management and contributing to the well-being of both the population and the environment in the region.

Acknowledgements

Acknowledgements go to Fardin Al Fahad Shakib, undergraduate pharmacy student, North South University, Bangladesh.

Conflict of interest

There is no conflict of interest.

Author details

Mohammad Borhan Uddin*, Ratin Al Samir and G.M. Golam Mehebub Pantho
Department of Pharmaceutical Sciences, School of Health and Life Sciences, North South University, Dhaka, Bangladesh

*Address all correspondence to: mohammad.uddin@northsouth.edu; bnn.nsu@gmail.com

References

[1] Abahussain E, Waheedi M, Koshy S. Practice, awareness and opinion of pharmacists toward disposal of unwanted medications in Kuwait. Saudi Pharmaceutical Journal. 2012;**20**(3):195-201

[2] Sarraf D, Rauniar G, Kushwaha R, Keshwar S, Maharjan R. Knowledge, attitude and practice of disposal of unused, unwanted and expired medicines among healthcare professionals. Kathmandu University Medical Journal. 2022;**20**(3):323-329

[3] Tong AY, Peake BM, Braund R. Disposal practices for unused medications around the world. Environment International. 2011;**37**(1):292-298

[4] Shakib FAF, Sadat N, Ahmed S, Nipa NY, Rahman M, Uddin MB. Unused and expired drug disposal practice and awareness among undergraduate students from pharmacy and other disciplines: Bangladesh perspective. Pharmacy Education. 2022;**22**(1):573-583

[5] Bashaar M, Thawani V, Hassali MA, Saleem F. Disposal practices of unused and expired pharmaceuticals among general public in Kabul. BMC Public Health. 2017;**17**:1-8

[6] Brechtelsbauer E, Shah S. Update on pharmaceutical waste disposal regulations: Strategies for success. American Journal of Health-System Pharmacy. 2020;**77**(7):574-582

[7] Alnahas F, Yeboah P, Fliedel L, Abdin AY, Alhareth K. Expired medication: Societal, regulatory and ethical aspects of a wasted opportunity. International Journal of Environmental Research and Public Health. 2020;**17**(3):787

[8] Hoque M, Rafi IK. Practice and awareness about unused and expired drug disposal among village people and city people in Bangladesh. GSC Biological and Pharmaceutical Sciences. 2023;**24**(3):132-139

[9] Freitas LAA, Radis-Baptista G. Pharmaceutical pollution and disposal of expired, unused, and unwanted medicines in the Brazilian context. Journal of Xenobiotics. 2021;**11**(2):61-76

[10] Ayele Y, Mamu M. Assessment of knowledge, attitude and practice towards disposal of unused and expired pharmaceuticals among community in Harar city, Eastern Ethiopia. Journal of Pharmaceutical Policy and Practice. 2018;**11**(1):27

[11] Angi'enda SA, Bukachi SA. Household knowledge and perceptions on disposal practices of unused medicines in Kenya. Journal of Anthropological Archaeology. 2016;**4**(2):1-20

[12] Kampamba M, Kalambwa Z, Chabalenge B, Zulu J, Mudenda S, Chimombe T, et al. Patients' knowledge, attitude and practices on disposal methods of expired and unused medicines: Implication for creation of drug take-back program. Pharmacology & Pharmacy. 2024;**15**(04):113-128

[13] Kamal KM, Chiumente M, Nakagawa S, Giannetti V, Marlin T. Disposal practices for unused and expired medications: Pilot data from three cities in three countries. GMS Health Innovation and Technologies. 2022;**16**

[14] Aditya S, Singh H. Safe medication disposal: Need to sensitize undergraduate students. International Journal of Pharmacy & Life Sciences. 2013;**4**(3)

[15] Watkins S, Barnett J, Standage M, Kasprzyk-HordernB,BardenR.Household disposal of pharmaceuticals: Attitudes and risk perception in a UK sample. Journal of Material Cycles and Waste Management. 2022;**24**(6):2455-2469

[16] de Campos EAR, Ten Caten CS, de Paula IC. End-of-use and end-of-life medicines—Insights from pharmaceutical care process into waste medicines management. Environmental Science and Pollution Research. 2021;**28**(41):58170-58188

[17] Kongar E, Haznedaroglu E, Abdelghany O, Bahtiyar MO. A novel IT infrastructure for reverse logistics operations of end-of-life pharmaceutical products. Information Technology and Management. 2015;**16**:51-65

[18] Nyaga MN, Nyagah DM, Njagi A. Pharmaceutical Waste: Overview, Management, and Impact of Improper Disposal. 2020

[19] Mwita S, Njegere M. Disposal practices of unfit medicines in public health facilities and community pharmacies: A case of Nyamagana District in Mwanza, Tanzania. Open Access Library Journal. 2024;**11**(6):1-9

[20] Daughton CG. Cradle-to-cradle stewardship of drugs for minimizing their environmental disposition while promoting human health. II. Drug disposal, waste reduction, and future directions. Environmental Health Perspectives. 2003;**111**(5):775-785

[21] Alghadeer S, Al-Arifi MN. Community pharmacists' practice, awareness, and beliefs about drug disposal in Saudi Arabia. Healthcare. MDPI; 2021;**9**(7):823

[22] Alshehri D, Banjar H. Increasing awareness of proper disposal of unused and expired medication using a knowledge-based disposal management system. Journal of Environmental and Public Health. 2022;**2022**(1):1797440

[23] Shwetha N, Jha A. Knowledge and awareness regarding safe drug disposal system among general population of India. Journal of Pharmacovigilance. 2018;**6**(2):256

[24] Raja S, Mohapatra S, Kalaiselvi A, Rani RJ. Awareness and disposal practices of unused and expired medication among health care professionals and students in a tertiary care teaching hospital. Biomedical and Pharmacology Journal. 2018;**11**(4):2073-2078

[25] Tegegne AA, Genet G, Workie Limenh L, Yohannes L, Mohammed Seid A, Alemayehu TT, et al. Public awareness, knowledge, and attitude regarding proper disposal of unused medicines and associated factors in Gondar city, Northwest Ethiopia. Frontiers in Public Health. 2024;**12**:1372739

[26] Brown T. Implications for a Drug Take Back Program in the Veterans Affairs Healthcare System. 2014

[27] Luo Y, Reimers K. Emergence of an Information Infrastructure through Integrating Waste Drug Recycling, Medication Management, and Household Drug Management in China. 2019

[28] Nipa N, Ahmed S, Shahariar M, Rahman M, Haider B, Uddin M. Improper management of pharmaceutical waste in south and south-east Asian regions. Journal of Environmental Studies. 2017;**3**(1):7

[29] Somashekara S, Suraj B. Knowledge, attitude, and practice of unused and expired medication disposal among second year medical students. National Journal of Physiology, Pharmacy and Pharmacology. 2023;**13**(3):561

[30] Asadullah K, Karthik G, Dharmappa B. A study on knowledge, attitude and practices regarding biomedical waste management among nursing staff in private hospitals in Udupi city, Karnataka, India. Revista Internacional de Geología, Ciencias Terrestres y Ambientales. 2013;**3**(1):118-123

[31] Azmi Hassali M, Shakeel S. Unused and expired medications disposal practices among the general public in Selangor, Malaysia. Pharmacy. 2020;**8**(4):196

[32] Manaia CM, Vaz-Moreira I, Nunes OC. Antibiotic resistance in waste water and surface water and human health implications. Emerging Organic Contaminants and Human Health. 2012:173-212

[33] Kachhap S, Singh GK, Soreng EA, Beck RK, Kumar U. A Study to Assess the Knowledge and Awareness Regarding Safe Disposal of Pharmaceutical Wastes among 2nd Year Medical and Para-Medical Students in a Tertiary Care Teaching Hospital in North India. 2020

[34] Manocha S, Suranagi UD, Sah RK, Chandane RD, Kulhare S, Goyal N, et al. Current disposal practices of unused and expired medicines among general public in Delhi and National Capital Region, India. Current Drug Safety. 2020;**15**(1):13-19

[35] Begum MM, Rivu SF, Hasan MMA, Nova TT, Rahman MM, Alim MA, et al. Disposal practices of unused and leftover medicines in the households of Dhaka metropolis. Pharmacy. 2021;**9**(2):103

[36] Al-Naggar RA, Alareefi A. Patients' opinion and practice toward unused medication disposal in Malaysia: A qualitative study. The Thai Journal of Pharmaceutical Sciences. 2010;**34**(3):117-123

[37] Azad MAK, Akhtaruzzaman MRHAM, Al-Mamun SM, Uddin M, Rahman M. Disposal practice for unused medications among the students of the international Islamic University Malaysia. Journal of Applied Pharmaceutical Science. 2012;**2**(7):101-106

[38] Avalee M, Hassan H. Disposal practice of unused medications among parents of school aged children in Rawang Selangor, Malaysia. Journal of Pharmacology and Toxicology Invest. 2015;**1**:22-26

[39] Insani WN, Qonita NA, Jannah SS, Nuraliyah NM, Supadmi W, Gatera VA, et al. Improper disposal practice of unused and expired pharmaceutical products in Indonesian households. Heliyon. 2020;**6**(7)

[40] Ahmed A, Mushtaq N. Disposal practices of unused and expired pharmaceuticals in Karachi and their impact on health and environment. Journal of University Medical & Dental College. 2013;**4**(2):42-48

[41] Arkaravichien W, Ruchipiyarak T, Thawinwan W, Benjawilaikul S. A threat to the environment from practice of drug disposal in Thailand. EnvironmentAsia. 2014;**7**(1)

[42] Zenker A, Cicero MR, Prestinaci F, Bottoni P, Carere M. Bioaccumulation and biomagnification potential of pharmaceuticals with a focus to

the aquatic environment. Journal of Environmental Management. 2014;**133**:378-387

[43] Phillips J. Pharmaceuticals in the Environment: Health Implications and Environmental Toxicity. Wayne State University; 2021

[44] Kharasch ED, Clark JD, Adams JM. Opioids and public health: The prescription opioid ecosystem and need for improved management. Anesthesiology. 2022;**136**(1):10-30

[45] Samal K, Mahapatra S, Ali MH. Pharmaceutical wastewater as emerging contaminants (EC): Treatment technologies, impact on environment and human health. Energy Nexus. 2022;**6**:100076

[46] Simu GM, Atchana J, Soica CM, Coricovac DE, Simu SC, Dehelean CA. Pharmaceutical mixtures: Still a concern for human and environmental health. Current Medicinal Chemistry. 2020;**27**(1):121-153

[47] Islam MM. Bacterial resistance to antibiotics: Access, excess, and awareness in Bangladesh. Expert Review of Anti-Infective Therapy. 2021;**19**(8):973-981

[48] Chaturvedi P, Shukla P, Giri BS, Chowdhary P, Chandra R, Gupta P, et al. Prevalence and hazardous impact of pharmaceutical and personal care products and antibiotics in environment: A review on emerging contaminants. Environmental Research. 2021;**194**:110664

[49] Daughton C. Pharmaceuticals as environmental pollutants: The ramifications for human exposure. International Encyclopedia of Public Health. 2008;**5**:66-102

[50] Nyagah D, Njagi A, Nyaga M. Expired drugs: Statistics, factors and mitigation strategies of med-gate pharmaceuticals and tan pharmacy up to 31/07/2021. Journal of Advanced Education and Sciences. 2022;**2**(2):09-11

[51] Kallaos J, Wheeler K, Wong C, Zahller M, Kotchen M. Pharmaceuticals in Wastewater Streams: Disposal Practices and Policy Options in Santa Barbara. Donald Bren School of environmental science and management; 2007

[52] Daughton CG. Cradle-to-cradle stewardship of drugs for minimizing their environmental disposition while promoting human health. I. Rationale for and avenues toward a green pharmacy. Environmental Health Perspectives. 2003;**111**(5):757-774

Chapter 12

Sustainable Management and Reuse of Sludge from Membrane Bioreactor Systems: Opportunities for Agricultural Application

Iva Ćurić, Ivan Brandić, Luka Brezinščak and Davor Dolar

Abstract

This chapter deals with the treatment and management of sludge resulting from wastewater treatment using membrane bioreactor (MBR) technology. With the increase in world population and industrial activities, the demand for effective wastewater treatment has increased, generating significant amounts of sludge. Although MBR technology is more efficient in terms of effluent quality, it produces sludge with different rheological and morphological characteristics than conventional activated sludge. This chapter discusses various methods for treating MBR sludge, including stabilization, dewatering, and thermal treatment, which are essential for improving sludge quality and reducing disposal costs. It also examines the potential for the reuse of MBR sludge in agriculture, highlighting methods such as composting and pelletization as sustainable approaches. These methods not only mitigate environmental impacts but also contribute to a circular economy by converting waste into valuable resources. The chapter concludes by highlighting the importance of proper sludge treatment and management to ensure compliance with environmental regulations and to fully utilize the potential of MBR sludge as a resource.

Keywords: membrane bioreactor, fertilizer, soil container, agriculture, sludge

1. Introduction

According to forecasts, the world's population will continue to grow over the next 50–60 years, reaching an estimated peak of around 10.3 billion people by the mid-2080s, up from 8.2 billion in 2024 [1, 2]. As the population grows, the demand for water will also increase significantly. In addition, this population growth often leads to an increase in industrial activity as the need for more goods, services, and infrastructure increases. The expansion of industry to meet these demands increases resource consumption, particularly water consumption, and leads to greater challenges in wastewater treatment and management [3].

Currently, most wastewater treatment plants rely on a combination of primary, secondary, and tertiary treatment processes to effectively treat wastewater. The main objectives of primary treatment include the removal of coarse and fine particles by screening, filtration, etc. Secondary treatment consists of biological processes, including aerobic and anaerobic processes, which are very effective in removing organic compounds and nutrients. Tertiary treatment, the final stage, involves chemical processes to further purify the wastewater [4]. All of the above processes produce a form of waste, namely sludge. Fytili and Zabaniotou state that on average, about 90 g of sludge is produced per person per day [5].

Although biological technologies such as conventional activated sludge (CAS) are very efficient at removing pollutants, they produce the most sewage sludge compared to all previous wastewater treatment processes [6]. Sewage sludge is the residual byproduct of wastewater treatment, which results in the separation of liquids and solids. The liquid fraction is discharged into water bodies, while the solid fraction undergoes further treatment and is ultimately disposed of. In the European Union, the USA, and China, between 18 and 33 million tons of dry weight are produced annually [6].

In 1969, as part of the Dorr-Oliver research program, Smith et al. presented a membrane bioreactor (MBR) for the first time and demonstrated its advantages over CAS technology, which enables higher effluent quality. MBR technology combines CAS with membrane separation using microfiltration, ultrafiltration, or nanofiltration membranes [7]. This combination offers several significant advantages compared to CAS. Nearly complete separation of sludge from the effluent and lower sludge production due to a higher solids retention time (SRT), a longer hydraulic retention time (HRT), and a higher concentration of mixed liquor suspended solids (MLSS) [8]. CAS and MBR technologies are presented in **Figure 1**.

However, the first MBR versions had to contend with considerable problems, in particular frequent fouling of the membranes and required costly maintenance [7]. In addition, these early systems were associated with high energy consumption, which further increased operating costs. Over the years, the MBR configuration has changed

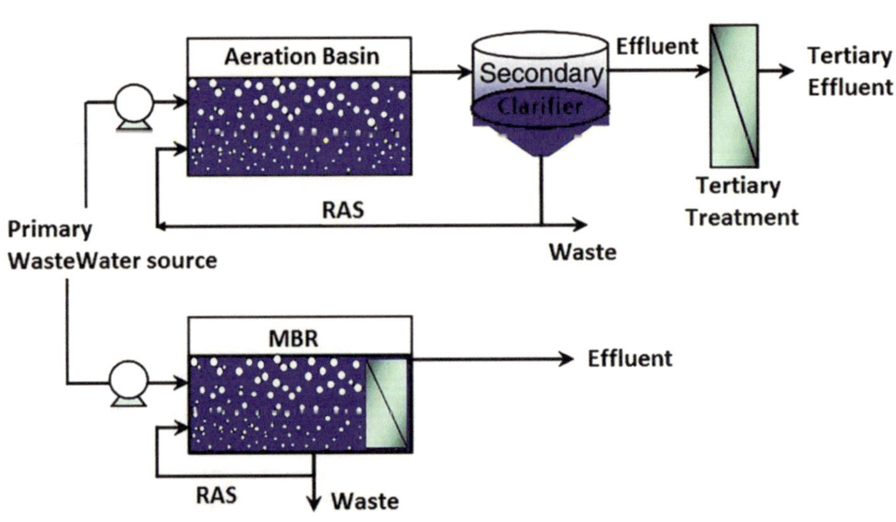

Figure 1.
MBR and CAS treatment of wastewater [9].

and overcome the abovementioned shortcomings. A clear indicator of the success and increasing acceptance of MBR technology is the global membrane bioreactor market, which was estimated at USD$ 3.35 billion in 2022 and is expected to grow at a compound annual growth rate (CAGR) of 7.0% from 2023 to 2030, highlighting the continuous improvement and wide acceptance of the technology [10].

Despite these advantages, both technologies generate sludge that has to be transported and treated. In Europe, however, the cost of sludge disposal can be as high as EUR 310 per ton, which can account for up to 65% of the total wastewater treatment costs [6]. One of the most popular methods of sludge treatment is incineration, which not only reduces the volume of sludge but also offers the possibility of generating thermal energy. However, this method has several significant disadvantages. Incineration requires a significant amount of energy to reach and maintain the high temperatures required for the process. It can also result in emissions of harmful gases such as CO_2, NO_x, and SO_x, which contribute to air pollution. The process also produces ash, which may contain heavy metals and other substances, requiring further treatment or disposal. Public acceptance can be a problem, as the process may meet with resistance due to potential environmental impacts and esthetic concerns. In addition, the high costs of technical cleaning and ash disposal add to the financial burden. Strict control measures are required to monitor and control pollutants and ensure compliance with environmental regulations [11].

Given these challenges, research into alternative approaches to sludge management is crucial. One such approach is the reuse of sludge, which not only mitigates some of the disadvantages associated with incineration but also offers economic benefits. Replacing conventional CAS technologies with MBR can also have a positive impact on sludge treatment. In addition, MBR sludge often has better dewatering characteristics and lower pathogen content compared to CAS sludge, making it easier and cheaper to process [12]. The Technical Report on End-of-Waste Criteria for Biodegradable Waste Subject to Biological Treatment classifies sewage sludge as a positive waste product, so that "clean" sewage sludge can be used as a fertilizer and is classified as a product and not as waste [13]. By reusing sewage sludge in this way, wastewater treatment plants can reduce their operating costs and even generate additional revenue. This approach is in line with the principles of the circular economy, where waste is transformed into valuable resources, contributing to both economic and environmental sustainability.

2. Characteristics of MBR sludge compared to CAS sludge

MBR sludge differs from CAS sludges not only quantitatively but also qualitatively, especially in terms of their rheology and morphology. These characteristics have a significant impact on system performance and the operating costs of the subsequent treatment process [14].

Sludge rheology includes viscosity, shear stress, shear rate, yield stress, and thixotropy. In contrast to CAS, MBR has a higher solids concentration, which has a major influence on sludge viscosity. The higher concentration of small flocs and the higher solids content in MBR sludge lead to higher viscosity. This makes the MBR sludge less flowable and makes effective circulation and mixing of the sludge more difficult. Mixing and circulating thick sludge requires a lot of energy. In addition, thick sludge contains much more water than CAS sludge, which also consumes more energy for the dewatering process, which is a kind of pretreatment before the actual sludge treatment [15]. Shear stress, shear rate, yield stress, and thixotropy describe the forces and behaviors required to initiate and maintain sludge movement, with the increased density of MBR sludge complicating these processes due to its greater resistance to flow and deformation. Pollice et al. [16] show in their study that with a higher concentration of suspended solids in the MBR, the energy required for mixing increases.

The sludge morphology describes the floc size, whereby the MBR flocs are less aggregated and therefore have a smaller particle size than in the conventional CAS system (**Figure 2**) [18]. Fenu et al. evaluated CAS and MBR sludge by microscopy. They found that the MBR flocs had open form and compact cores, cell clusters, filamentous bacteria, protozoa, metazoa, algae, and fungi. The particle size was 40–50 μm for the MBR and 200 μm for the CAS [19].

The microorganisms mentioned form a biofilm of extracellular polymeric substances (EPS). These substances play a key role in the aggregation of flocs in MBR sludge and contribute to the formation of dense and compact structures. In CAS sludge, this aggregation is less pronounced due to the lower EPS concentration. The EPS concentration is related to the SRT. A longer SRT in the MBR leads to a higher EPS concentration [20]. According to Pontoni et al. [12], CAS EPS consists mainly of proteins (57–61%), carbohydrates (28–29%), humic acids (6–7%), and uronic acids (4–8%). On the other hand, MBR sludge contains lower proportions of proteins (17–50%), carbohydrates (13–22%), humic acids (2–4%), and uronic acids (32–60%).

Overall, EPS affects the morphology of the sludge and has a significant impact on the stability of the sludge, the degradability of the substrate, and other hydrodynamic conditions.

In addition to the rheology and morphological properties of the sludge, it is necessary to know the chemical composition of the sludge before treatment and disposal. However, this depends on the type of water in question, that is, the source from which it originates. For example, if the wastewater comes from the metal industry, it is likely to have a high concentration of heavy metals. This information is important if the sludge can be disposed of in a landfill [5]. Without proper treatment, disposal of sludge has catastrophic consequences for the environment. Alonso Álvarez et al. conducted a study on the sequential extraction of metals (Al, Cd, Co, Cu, Cr, Fe, Mn, Hg, Mo, Ni, Pb, Ti, and Zn) from sludge samples collected from five different municipal activated sludge plants. The study found that Cd, Co, Mo, Ni, and Ti pose only a low toxic risk due to their low concentrations, while Cr, Cu, and Pb were present in higher concentrations, which requires further investigation to assess their potentially harmful effects on the biological processes in the sludge [21]

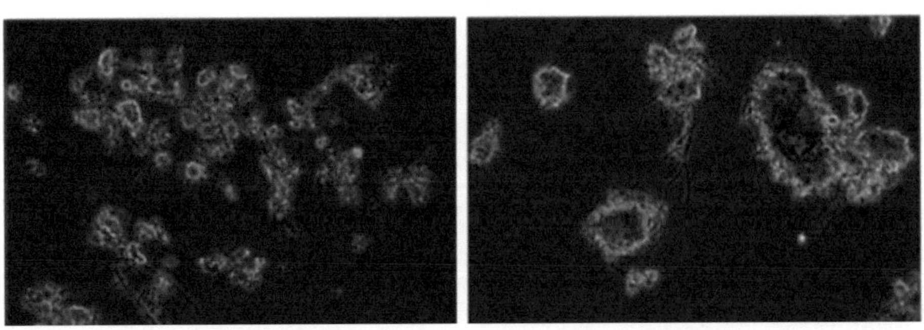

Figure 2.
CAS sludge flocs (left) and MBR sludge flocs (right) [17].

3. Treatment of the MBR sludge

Sludge treatment involves several steps, including thickening to reduce volume by removing water, stabilization to reduce organic content, odors, and pathogens, and dewatering and drying to further remove water and facilitate handling. Finally, testing for contaminants such as heavy metals and pathogens ensures that safety standards for land application are met. These steps are similar in both CAS and MBR, but there are some differences [22].

Stabilization is applied to both, but MBR sludge may require additional measures due to its different composition, especially the higher content of uronic acids. Dewatering and drying processes include dewatering and drying, although MBR sludge may be more difficult to dewater due to its higher EPS content. Safety testing is required for both application, but MBR sludge may have unique characteristics due to the different microbial activity in the system [12].

The dewatering process reduces the content of bound water, which consists of three types of water: interstitial, vicinal, and hydration water. Interstitial water is located in the spaces between the solid particles of the sludge and is relatively easy to remove. Vicinal water is bound to the surfaces of the sludge particles and is more difficult to remove than interstitial water. The hydration water is chemically bound to the sludge particles and is the most difficult to remove. Belt processes, centrifuges, and filter processes can press the water out of the MBR sludge [23]. The dewaterability process is very difficult due to the particle size of MBR sludge, but a longer SRT can solve this problem. D'Antonio and Napoli investigated the dewaterability of chromium-containing sludge from the tannery wastewater treatment plant. The experiment shows that the MBR sludge has a specific resistance to filtration (SRF) of 2.58×10^{11} m/kg and an SRT of 25 days. The results show better dewaterability, as MBR sludge contains more free water in this case [24].

Stabilization of MBR sludge is essential to improve its quality and handling. Methods include anaerobic and aerobic digestion and thermal stabilization [17].

Aerobic digestion of MBR sludge is a process in which microorganisms break down organic matter in the presence of oxygen. This process is extremely efficient in reducing sludge volume and eliminating pathogens. In aerobic digestion, microorganisms use dissolved oxygen to break down complex organic molecules, resulting in the release of carbon dioxide, water, and heat. The process is fast and produces a stable sludge with lower levels of pathogens. However, aerobic digestion is energy-intensive as it requires constant aeration, which can increase operating costs [25].

On the other hand, anaerobic digestion is a process of decomposition of organic material in the absence of oxygen. This process is slower compared to aerobic digestion but has the great advantage of producing biogas, a mixture of methane and carbon dioxide that can be used as a renewable energy source. In anaerobic digestion, microorganisms first hydrolyze complex organic matter into simpler molecules, which are then fermented into acids. Finally, methanogenic bacteria convert these acids into methane and carbon dioxide. This process is very effective in reducing sludge volume and generating energy but requires careful management as the methanogenic bacteria are sensitive to changes in environmental conditions, such as pH and temperature [17, 25].

In combination with MBR systems, anaerobic digestion can reduce the volume of sludge produced and at the same time generate biogas as an additional resource. Due to the long SRT in MBR systems, the resulting sludge usually has a high concentration of anaerobically degradable substances, which can increase the efficiency of anaerobic digestion. In addition, anaerobic digestion can help reduce the operating costs associated with sludge treatment while contributing to sustainability through the production of renewable energy [17].

Thermal stabilization of MBR sludge is a process in which the sludge produced in MBR systems is treated with heat and stabilized.

Due to the unique characteristics of MBR sludge, which often contains higher concentrations of biomass and EPS, thermal stabilization can help to degrade these complex materials more effectively. The main methods of thermal stabilization of MBR sludge include thermophilic aerobic digestion, thermal hydrolysis, and in some cases, incineration.

1. Thermophilic aerobic digestion

   In this process, the MBR sludge is treated at elevated temperatures (usually 45–70°C) under aerobic conditions. The higher temperatures accelerate the decomposition of the organic matter, which reduces the sludge volume more effectively and destroys pathogens. MBR sludge, which is already partially stabilized due to the long SRT, responds well to thermophilic conditions, resulting in faster stabilization and greater reduction of volatile solids.

2. Thermal hydrolysis

   Thermal hydrolysis can be particularly beneficial for MBR sludge. In this process, the sludge is exposed to high temperatures (usually 160–180°C) and pressures before anaerobic digestion. The intense conditions break down complex organic compounds and EPS, making the sludge more biodegradable. When MBR sludge is subjected to thermal hydrolysis, it is not only easier to anaerobically digest, but the subsequent digestion also produces more biogas, which contributes to energy production.

3. Incineration

   Although the incineration of MBR sludge is less common, it can still be used, especially if energy recovery is the main objective. Since MBR sludge has a high organic content, it can be effectively reduced to ash through incineration while recovering energy in the form of heat. However, due to the typically lower water content of MBR sludge compared to conventional sludge, pre-drying may be required to make incineration more efficient.

Thermal hydrolysis is an energy-intensive process that can significantly increase the carbon footprint of sludge management. While biogas production of biogas can offset some of these energy costs, the overall energy balance remains a concern, particularly for systems that rely heavily on nonrenewable energy sources [26]. Furthermore, while incineration can reduce sludge volume and generate thermal energy, it is also criticized for high energy requirements and the emission of harmful gases such as CO_2, NO_x, and SO_x. These emissions contribute to air pollution and climate change and require strict control measures and advanced technological solutions to mitigate their effects.

4. Agricultural benefit of MBR sludge

The conversion of MBR sludge into fertilizer or other useful products not only reduces disposal costs but also contributes to the circular economy and sustainable use of resources. MBR sludge usually contains significant amounts of organic matter, nitrogen, phosphorus, and trace elements such as potassium, calcium, magnesium, and sulfur. These substances are major components of most commercial fertilizers, making MBR sludge a potentially valuable source of nutrients for agricultural applications [27]. However, before application, the sludge must be treated as described in the previous section.

1. Composting of the MBR sludge

   Composting is one of the most widely used methods for converting stabilized MBR sludge into valuable fertilizer, mainly due to its compliance with the principles of green technology. Composting solves the problems associated with sludge through controlled aerobic fermentation, which effectively converts organic matter into stable humic compounds. This process not only reduces pathogens but also results in nutrient-rich humus, which is an excellent, high-quality organic fertilizer. In composting, stabilized sludge is usually mixed with organic materials such as sawdust, food waste, or agricultural residues to enhance the process. Effective aeration is crucial, but dewatered sludge often has low porosity. To improve aeration, a bulking agent, such as a carbonaceous material, is added and at least 20% of the mixture must be permeable to air. This controlled degradation, which is driven by various microorganisms and generates heat that can exceed 70°C, effectively destroying pathogenic germs and reducing the moisture content of the sludge through evaporation. Composting is most effective with fresh sludge rich in organic matter and nitrogen, but can also be used with digested or aerobically stabilized sludge. The process is naturally supported by the microflora present in the sludge or in the air, so no additional seed is required. The pH of the mixture is self-regulating between 6.5 and 8, making composting a robust method even for sludge produced from physical-chemical treatments. Due to its effectiveness and consistency with sustainable development goals, composting has become an increasingly popular solution for the management and reuse of sludge in recent years [28, 29].

2. Pelletization of MBR sludge

   MBR sludge pelletization is a process that transforms stabilized sludge into a more manageable form so it is easier to handle, transport, and use in agriculture or other industries. This method involves several important steps, each of which contributes to the formation of small, solid pellets that are rich in nutrients [30]. The pelletization process includes several processes:

   a. Preparation and drying:

      In the first phase of pelletization, the moisture content of the sludge is reduced. Drying is crucial, as a high moisture content can hinder the pelletization process and affect the quality of the final product. Drying can be carried out by thermal methods or by natural drying, depending on the available resources and the specific requirements of the process [31].

b. Grinding and homogenization

After drying, the sludge is ground to achieve a uniform consistency. This step is important to ensure that the sludge is pelletized evenly, resulting in pellets of uniform size and density. Homogenization also allows the sludge to be effectively mixed with additional materials, such as binding agents, which can be added to improve the properties of the pellets [32].

c. Pelletization

The ground and dried sludge is then fed into a pellet mill, where it is compressed into solid, cylindrical pellets. In the pellet mill, the sludge is formed into pellets under pressure and heat, which are then cooled to maintain their shape and structure. Sometimes, binding agents are added during this process to improve the strength and durability of the pellets [31].

d. Cooling and packaging

Once the pellets are formed, they must be cooled before they can be stored or transported. Cooling prevents the pellets from breaking apart and ensures that they retain their structural integrity. Once cooled, the pellets are packaged in bags or other suitable containers so they can be distributed and used [7].

3. Direct application into the field as a soil conditioner

A soil conditioner, also known as a soil amendment or improver, is a substance that is added to soil to improve its physical properties, such as its structure, porosity, and water retention capacity. Soil conditioners aim to improve the overall health and productivity of the soil by modifying its physical qualities [33].

The use of MBR sludge as a soil conditioner involves several steps that contribute to improving soil quality:

a. Preparation and drying of sludge

First, the MBR sludge must be properly treated and dried. Drying is crucial as it reduces the moisture content of the sludge, improves its stability, and facilitates its handling.

b. Application of sludge to soil

The dried sludge is spread evenly over the soil surface. The amount of sludge applied depends on the specific needs of the soil and the type of soil to be treated.

c. Incorporation into the soil

Once the sludge has been spread, it must be incorporated into the top layer of soil. This is usually done by plowing or tilling. This ensures that the sludge is well mixed with the existing soil, which helps to improve soil structure and increase porosity [34].

Overall, the incorporation of MBR sludge into the soil can significantly improve its physical properties, resulting in better soil structure, higher water retention capacity, and reduced soil erosion. The organic matter in MBR sludge binds soil particles together and creates a looser, more workable soil that supports vigorous root growth. In addition, the improved water retention is particularly beneficial for sandy or dry soils, while the improved structure prevents erosion and protects the topsoil. Although MBR sludge is not a fertilizer in the traditional sense, it contributes to soil fertility, which can ultimately lead to higher agricultural productivity.

5. General challenges in the reuse of MBR sludge

Despite the numerous benefits of MBR sludge reuse, such as improved dewatering properties and reduced pathogen content, significant problems remain. One of the main problems is the presence of emerging contaminants, such as pharmaceutical residues, personal care products, and microplastics, which are not always completely removed by conventional treatment processes. These substances can accumulate in the soil over time and pose a potential risk to human health, the environment, and long-term soil fertility. Heavy metal contamination also remains a key issue, as toxic metals can gradually accumulate when sewage sludge is applied to agricultural land, which could affect crop safety and quality [35].

The logistical challenges of transporting and spreading sewage sludge are also significant, especially when it comes to large-scale agricultural applications. The cost of sludge transportation can be prohibitive, especially in regions where agricultural land is far from wastewater treatment plants [36]. In addition, the varying composition of sludge depending on the source of the wastewater makes it difficult to apply uniformly. MBR sludge in particular can have significant differences in nutrient and pollutant content, which requires careful monitoring and testing before it can be safely reused.

Another challenge is the legal framework for the reuse of sewage sludge. While some regions have established guidelines for the safe use of treated sewage sludge in agriculture, in other regions there are no clear regulations, leading to uncertainty and hesitation among potential users. Even in regions with strict standards, it can be a lengthy and costly process for wastewater treatment plants to obtain the necessary permits and certifications [37]. Public perception also plays an important role, as the reuse of sludge in agriculture is often viewed with skepticism. Concerns about unpleasant odors, contamination, and the possibility of pathogens or chemicals entering the food chain can lead to resistance from farmers and consumers alike. To overcome these concerns, extensive awareness campaigns and transparent communication about the safety measures involved in sludge treatment are needed.

Finally, the long-term effects of MBR sludge application on soil health are not yet fully known. While the short-term benefits of organic matter, nitrogen and phosphorus enrichment are obvious, the cumulative effects of repeated sludge applications need to be further investigated. Problems such as soil compaction, changes in microbial communities and nutrient imbalances could occur over time, so continued research is needed to ensure that the practice remains sustainable in the long term.

6. Possible future innovative sludge treatment methods

Current research is exploring a range of advanced technologies aimed at improving the efficiency of sludge treatment and reducing the environmental impact. For example, advanced oxidation processes are being investigated for their ability to degrade persistent organic pollutants and microplastics that are difficult to remove using conventional methods [38]. Hydrodynamic cavitation is another promising technique that supports sludge degradation, reduces particle size, and improves stabilization. This process, which is suitable for both anaerobic and aerobic digestion systems, facilitates the handling and reuse of sewage sludge. The process of stabilizing sewage sludge with materials such as clay minerals and biochar provides an effective way to reduce microbial activity, manage heavy metals, and optimize nutrient recycling. Studies suggest that this method can be used to produce a safe soil amendment and fertilizer that is an environmentally friendly option for agriculture [39].

Another promising area of innovation is the integration of artificial intelligence (AI) and machine learning to optimize sludge treatment processes. AI-driven systems can analyze large data sets to predict and adjust energy consumption, chemical dosing, and system parameters in real-time, leading to a significant reduction in energy consumption and operating costs [40]. In addition, waste-to-energy technologies continue to evolve, with research focusing on maximizing biogas production through anaerobic digestion or exploring pyrolysis and gasification as alternatives to conventional incineration [41]. These processes not only reduce the volume of sludge but also convert it into syngas, biochar, or other valuable by-products, further contributing to the circular economy.

7. Conclusion

The use of MBR sludge for agricultural reclamation offers a sustainable approach to wastewater treatment that is in line with the principles of the circular economy. As the world's population continues to grow, the demand for water and the need for efficient wastewater management is increasing. MBR technology offers several advantages over CAS technology, including better effluent quality and reduced sludge production. However, the treatment and disposal of MBR sludge remains a challenge due to its unique properties, such as higher viscosity and solids content, which require more energy for treatment.

By reusing MBR sludge as a resource for agricultural reclamation, we can address these challenges while contributing to sustainability. MBR sludge, which is rich in organic matter, nitrogen, phosphorus, and trace elements, has significant potential as a soil conditioner or fertilizer. Methods such as composting, palletization, and direct soil application turn this byproduct into a valuable resource for improving soil structure, fertility, and water retention. In addition, incorporating treated MBR sludge into agricultural practices reduces disposal costs, mitigates environmental impacts, and promotes the recovery of renewable resources through processes such as biogas production.

Thus, the reuse of MBR sludge not only improves the sustainability of wastewater treatment but also supports agricultural productivity and offers a holistic solution for resource management in an increasingly resource-constrained world.

Acknowledgements

This investigation was financed by the NATO Science for Peace and Security Program under Grant ID. G6087.

Conflict of interest

The authors declare no conflict of interest.

Notes/thanks/other declarations

I would like to express my sincere gratitude to all the authors who have contributed to this chapter. Your invaluable insights, research, and expertise have greatly enriched the content and ensured a comprehensive understanding of the topic. I greatly appreciate your cooperation and commitment and am grateful for the time and effort you have invested in this work. Thank you for your outstanding contributions.

Author details

Iva Ćurić[1], Ivan Brandić[2], Luka Brezinščak[2] and Davor Dolar[1*]

1 University of Zagreb Faculty of Chemical Engineering and Technology, Zagreb, Croatia

2 University of Zagreb Faculty of Agriculture, Zagreb, Croatia

*Address all correspondence to: dolar@fkit.unizg.hr

References

[1] Seleiman MF, Santanen A, Mäkelä PSA. Recycling sludge on cropland as fertilizer—Advantages and risks. Vol. 155. New York: Resources, Conservation and Recycling; 2020. p. 104647. DOI: 10.1016/j.resconrec.2019.104647

[2] DESA/Population Division. World Population Prospects 2024: Summary of Results. United Nations Department of Economic and Social Affairs, Population Division; 2024. p. 67

[3] Wang J, Yang Z, Qian X. Driving factors of urban shrinkage: Examining the role of local industrial diversity. Cities. 2020;**99**:102646. DOI: 10.1016/j.cities.2020.102646

[4] Ding GKC. Wastewater treatment, reused and recycling—A potential source of water supply. In: Abraham MA, editor. Encyclopedia of Sustainable Technologies. 2nd ed. Oxford: Elsevier; 2024. pp. 676-693

[5] Fytili D, Zabaniotou A. Utilization of sewage sludge in EU application of old and new methods—A review. Renewable and Sustainable Energy Reviews. 2008;**12**:116-140. DOI: 10.1016/j.rser.2006.05.014

[6] Ferrentino R, Langone M, Fiori L, Andreottola G. Full-scale sewage sludge reduction technologies: A review with a focus on energy consumption. Water. 2023;**15**:615

[7] Al-Asheh S, Bagheri M, Aidan A. Membrane bioreactor for wastewater treatment: A review. Case Studies in Chemical and Environmental Engineering. 2021;**4**:100109. DOI: 10.1016/j.cscee.2021.100109

[8] Qrenawi LI, Rabah FKJ. Studying the effect of HRT, SRT, and MLSS on membrane bioreactor performance for wastewater treatment. Egyptian Journal of Chemistry. 2024;**67**:423-437. DOI: 10.21608/ejchem.2024.239415.8684

[9] Asante-Sackey D, Rathilal S, Tetteh EK, Armah EK. Membrane bioreactors for produced water treatment: A mini-review. Membranes 275. 2022;**12**:275

[10] I.B. Chemicals. Membrane Bioreactor Market Size, Share & Trends Analysis Report By Product (Hollow Fiber), By Configuration (Submerged), By Application (Municipal), By Region, And Segment Forecasts. California: Grand View Research, Inc.; 2023-2030. 2023. p. 125

[11] Zheng A, Li L, Tippayawong N, Huang Z, Zhao K, Wei G, et al. Reducing emission of NO_x and SO_x precursors while enhancing char production from pyrolysis of sewage sludge by torrefaction pretreatment. Energy. 2020;**192**:116620. DOI: 10.1016/j.energy.2019.116620

[12] Pontoni L, Papirio S, D'Alessandro G, Caniani D, Gori R, Mannina G, et al. Dewaterability of CAS and MBR sludge: Effect of biological stability and EPS composition. Journal of Environmental Engineering. 2018;**144**:04017088. DOI: 10.1061/(ASCE)EE.1943-7870.0001299

[13] European Commission: Joint Research Centre, Kowalska M, Delre A, Wolf O. EU Ecolabel Criteria for Growing Media and Soil Improvers. Luxembourg: Publications Office of the European Union; 2021. DOI: 10.2760/748007

[14] Tang B, Zhang Z. Essence of disposing the excess sludge and optimizing the operation of wastewater treatment: Rheological behavior and microbial ecosystem. Chemosphere.

2014;**105**:1-13. DOI: 10.1016/j.chemosphere.2013.12.067

[15] Yousefi SA, Nasser MS, Hussein IA, Benamor A, El-Naas MH. Influence of polyelectrolyte structure and type on the degree of flocculation and rheological behavior of industrial MBR sludge. Separation and Purification Technology. 2020;**233**:116001. DOI: 10.1016/j.seppur.2019.116001

[16] Pollice A, Giordano C, Laera G, Saturno D, Mininni G. Physical characteristics of the sludge in a complete retention membrane bioreactor. Water Research. 2007;**41**:1832-1840. DOI: 10.1016/j.watres.2007.01.022

[17] Campo R, Capodici M, Corsino F, Di Bella G, Esposito G, Fabbricino M, et al. 14—Membrane bioreactors sludge: From production to disposal. In: Mannina G, Pandey A, Larroche C, Ng HY, Ngo HH, editors. Current Developments in Biotechnology and Bioengineering. Amsterdam: Elsevier; 2020. pp. 323-351

[18] Lee SA, Fane AG, Amal R, Waite TD. The effect of floc size and structure on specific cake resistance and compressibility in dead-end microfiltration. Separation Science and Technology. 2003;**38**:869-887. DOI: 10.1081/SS-120017631

[19] Fenu A, Roels J, Wambecq T, De Gussem K, Thoeye C, De Gueldre G, et al. Energy audit of a full scale MBR system. Desalination. 2010;**262**:121-128. DOI: 10.1016/j.desal.2010.05.057

[20] Huang L, Jin Y, Zhou D, Liu L, Huang S, Zhao Y, et al. A review of the role of extracellular polymeric substances (EPS) in wastewater treatment systems. International Journal of Environmental Research and Public Health. 2022;**19**:1-14. DOI: 10.3390/ijerph191912191

[21] Álvarez EA, Mochón MC, Sánchez JCJ, M.T. Rodríguez. Heavy metal extractable forms in sludge from wastewater treatment plants. Chemosphere. 2002;**47**:765-775. DOI: 10.1016/S0045-6535(02)00021-8

[22] Kamizela T, Kowalczyk M. 18—Sludge dewatering: Processes for enhanced performance. In: Prasad MNV, de Campos Favas PJ, Vithanage M, Mohan SV, editors. Industrial and Municipal Sludge. Oxford: Butterworth-Heinemann; 2019. pp. 399-423

[23] Chen G, Lock Yue P, Mujumdar AS. Sludge dewatering and drying. Drying Technology. 2002;**20**:883-916. DOI: 10.1081/DRT-120003768

[24] d'Antonio L, Napoli RMA. Dewaterability of MBR sludge loaded with tannery wastewater. Desalination and Water Treatment. 2010;**23**:129-134. DOI: 10.5004/dwt.2010.1986

[25] Kazimierowicz J, Dębowski M. Aerobic granular sludge as a substrate in anaerobic digestion—Current status and perspectives. Sustainability. 2022;**14**:10904

[26] Singh DK, Garg A. Thermal hydrolysis of sewage sludge: Improvement in biogas generation and prediction of global warming potential. Waste Management & Research. 2023;**42**:51-58. DOI: 10.1177/0734242X231171044

[27] Gherghel A, Teodosiu C, De Gisi S. A review on wastewater sludge valorisation and its challenges in the context of circular economy. Journal of Cleaner Production. 2019;**228**:244-263. DOI: 10.1016/j.jclepro.2019.04.240

[28] Grgas D, Štefanac T, Barešić M, Toromanović M, Ibrahimpašić J, Vukušić Pavičić T, et al. Co-composting of sewage sludge, green waste, and food waste.

Journal of Sustainable Development of Energy, Water and Environment Systems. 2023;**11**:1-14

[29] Ayilara MS, Olanrewaju OS, Babalola OO, Odeyemi O. Waste management through composting: Challenges and potentials. Sustainability. 2020;**12**:4456

[30] Yilmaz E, Wzorek M, Akçay S. Co-pelletization of sewage sludge and agricultural wastes. Journal of Environmental Management. 2018;**216**:169-175. DOI: 10.1016/j.jenvman.2017.09.012

[31] Cusidó JA, Soriano C. Valorization of pellets from municipal WWTP sludge in lightweight clay ceramics. Waste Management. 2011;**31**:1372-1380. DOI: 10.1016/j.wasman.2011.02.003

[32] Siyal AA, Liu Y, Mao X, Siyal ZA, Ran C, Ao W, et al. Co-pelletization of sewage sludge, furfural residue and corn stalk: Characteristics and quality analysis of pellets. Biomass and Bioenergy. 2021;**150**:106121. DOI: 10.1016/j.biombioe.2021.106121

[33] Babla M, Katwal U, Yong M-T, Jahandari S, Rahme M, Chen Z-H, et al. Value-added products as soil conditioners for sustainable agriculture. Resources, Conservation and Recycling. 2022;**178**:106079. DOI: 10.1016/j.resconrec.2021.106079

[34] Ingelmo F, Canet R, Ibañez MA, Pomares F, García J. Use of MSW compost, dried sewage sludge and other wastes as partial substitutes for peat and soil. Bioresource Technology. 1998;**63**:123-129. DOI. 10.1016/S0960-8524(97)00105-3

[35] Achkir A, Aouragh A, El Mahi M, Lotfi EM, Labjar N, El Bouch M, et al. Implication of sewage sludge increased application rates on soil fertility and heavy metals contamination risk. Emerging Contaminants. 2023;**9**:100200. DOI: 10.1016/j.emcon.2022.100200

[36] Lue-Hing C, Matthews P, Nàmer J, Okuno N, Spinosa L. Sludge management in highly urbanized areas. Water Science and Technology. 1996;**34**:517-524. DOI: 10.1016/0273-1223(96)00619-1

[37] Spinosa L, Molinari L. Standardization: A necessary support for the utilization of sludge/biosolids in agriculture. Standards. 2023;**3**(4):385-399

[38] Gao Q, Li L, Zhang Y, Zhou H, Jiang J, Wei L, et al. Advanced oxidation processes (AOPs)-based sludge pretreatment techniques for enhanced short-chain fatty acids production: A critical review. Chemical Engineering Journal. 2024;**489**:151496. DOI: 10.1016/j.cej.2024.151496

[39] Massahi T, Hossini H, Parnoon K, Atashkar S, Sharafi M, Massahi D, et al. Future of sludge management. In: Kılıç TB, Eduardo J-L, Queiroz ZL, Costa DM, editors. Wastewater Treatment and Sludge Management Systems—The Gutter-to-Good Approaches. London, UK, Rijeka: IntechOpen; 2023. Ch. 9

[40] Miao S, Zhou C, AlQahtani SA, Alrashoud M, Ghoneim A, Lv Z. Applying machine learning in intelligent sewage treatment: A case study of chemical plant in sustainable cities. Sustainable Cities and Society. 2021;**72**:103009. DOI: 10.1016/j.scs.2021.103009

[41] Ronda A, Haro P, Gómez-Barea A. Sustainability assessment of alternative waste-to-energy technologies for the management of sewage sludge. Waste Management. 2023;**159**:52-62. DOI: 10.1016/j.wasman.2023.01.025